Elementary Astronomy

THE UNIVERSE

From Comets to Constellations

Teacher's Guide

Tom DeRosa
Carolyn Reeves

THE UNIVERSE

From Comets to Constellations

Teacher's Guide

First printing: April 2014

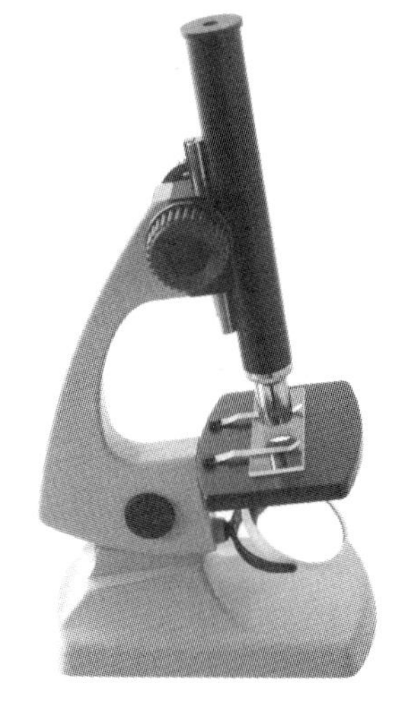

Master Books® is a division of the New Leaf Publishing Group, Inc.

ISBN: 978-0-89051-799-4

Cover by Diana Bogardus

Unless otherwise noted, Scripture quotations are from the New International Version of the Bible.

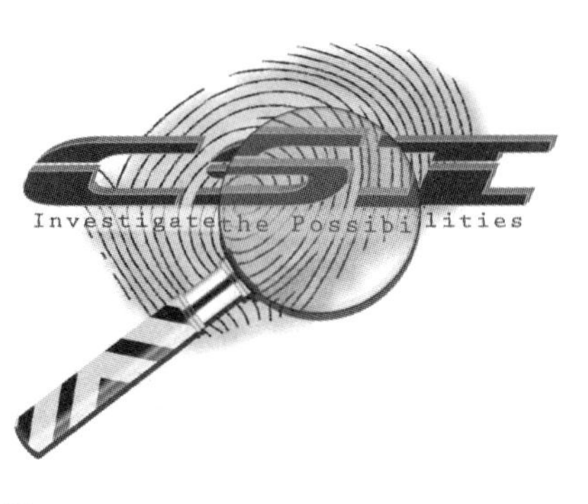

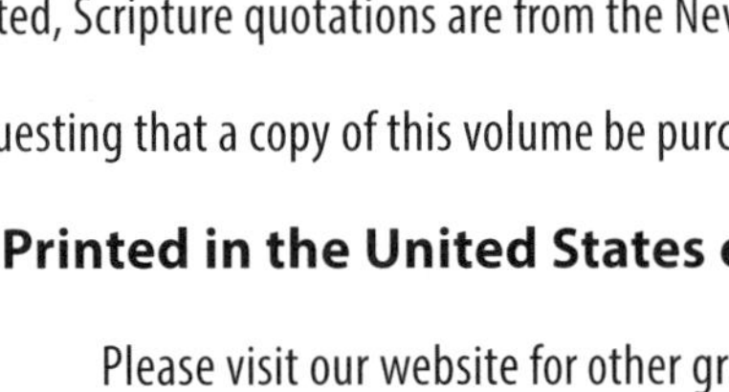

Please consider requesting that a copy of this volume be purchased by your local library system.

Printed in the United States of America

Please visit our website for other great titles:
www.masterbooks.net

For information regarding author interviews, please contact the publicity department at (870) 438-5288

Master Books®
A Division of New Leaf Publishing Group
www.masterbooks.net

Table of Contents

Introduction

Unless they have access to good telescopes, students may miss some of the most exciting parts of astronomy. Fortunately, there are several ways to take advantage of some of the new information about space. For example, NASA provides a wealth of information and photographs about space science that is available to students.

There are also a number of versions of the investigations in this book that can be seen on YouTube and other sites. Both teachers and students can be involved in searching for related investigations. Be sure to keep a record of helpful videos and websites that have been located. As you look for information, be aware that many Internet sites place a heavy emphasis on how the universe evolved over billions of years. The recommended references given in this book are excellent sources of information.

Look for ways to let students view the night sky with a good telescope. You might plan a trip around a visit to a nearby university or museum where telescope viewing if available. Watch the news for meteor showers, eclipses, and other unusual events in the sky. You might even consider investing in a telescope.

1. Think about This

The purpose of this section is to introduce something that will spark an interest in the upcoming investigation. Lesson beginnings are a good time to let students make observations on their own; for a demonstration by the teacher; or to include any other kind of engaging introduction that causes the students to want to get answers. Teachers should wait until after students have had an opportunity to do the investigation before answering too many questions. Ideally, lesson beginnings should stimulate the students' curiosity and make them want to know more. Lesson beginnings are also a good time for students to recall what they already know about the lesson topic. Making a connection to prior knowledge makes learning new ideas easier.

2. The Investigative Problem

This section brings a focus to the activity students are about to investigate and states the objectives of the lesson. Students should be encouraged throughout the investigation to ask questions about the things they want to know. It is the students' questions that connect with the students' natural curiosity and makes them want to learn more. Teachers should stress to students at the start of each lesson that the goal is to find possible solutions for the investigative problem.

3. Gather These Materials

All the supplies and materials that are needed for the investigation are listed. The teacher's book may contain additional information about substituting more inexpensive or easier to find materials.

4. Procedures and Observations

Instructions are given about how to do the investigation. The teacher's book may contain more specifics about the investigations. Students will write their observations as they perform the activity.

5. The Science Stuff

It is much easier for students to add new ideas to a topic in which they already have some knowledge or experience than it is to start from scratch on a topic they know nothing about. This section builds on the experience of the investigation.

6. Making Connections

Lessons learned become more permanent when they are related to other situations and ideas in the world. This section reminds students of concepts and ideas they likely already know. The scientific explanation for what the students observed should be more meaningful if it can be connected to other experiences and/or prior knowledge. The more connections that are made, the greater the students' level of understanding will become.

7. Dig Deeper

This section provides ideas for additional things to do or look up at home. Students will often want to learn more than what was in the lesson. This will give them some choices for further study. Students who show an interest in their own unanswered questions should be allowed to pursue their interests, provided the teacher approves of an alternative project. Students should aim to do at least one project per week from Dig Deeper or other project choices, recording the projects they choose to do, along with the completion date, in a notebook or journal. The minimum requirements from this section should correspond to each student's grade level.

8. What Did You Learn?

This section contains a brief assessment of the content of the lesson in the form of mostly short-answer questions.

9. The Stumpers Corner

The students may write two things they would like to learn more about or two "stumper" questions (with answers) pertaining to the lesson. Stumper questions are short-answer questions to ask to family or classmates, but they should be hard enough to be a challenge.

Note to the Teacher

The books in this series are designed to be applicable mainly for grades 3–8. The recommendations for K–4 were also considered, because basic content builds from one level to another. The built-in flexibility allows younger students to do many of the investigations, provided they have good reading and math skills. Middle school students will be presented the basic concepts for their level, but will benefit from doing more of the optional research and activities.

We feel it is best to leave grading up to the discretion of the teacher. However, for those who are not sure what would be a fair way to assess student work, the following is a suggestion.

1. Completion of 20 activities with write-up of observations — ⅓
2. Completion of What Did You Learn Questions + paper and pencil quizzes — ⅓
3. Projects, Contests, and Dig Deeper — ⅓

The teacher must set the standards for the amount of work to be completed. The basic lessons will provide a solid foundation for each unit, but additional research and activities are a part of the learning strategy. The number of required projects should depend on the age, maturity, and grade level of the students. All students should choose and complete at least one project each week or 20 per semester. 5th and 6th graders should complete 25 projects per semester. A minimum guide for 7th and 8th graders would be 30 projects. The projects can be chosen from "Dig Deeper" ideas or from any of the other projects and features. Additional projects would give extra credits. By all means, allow students to pursue their own interests and design their own research projects, as long as you approve first. Encourage older students to do the more difficult projects.

As students complete each investigation and other work, they should record what they did and the date completed in the student journal. You may or may not wish to assign a grade for total points. But a fair evaluation would be three levels, such as: minimum points, more than required, and super work. Remember, the teacher sets the standards for evaluating the work.

Ideally, if students miss one of the investigations, they should find time to make it up. When this is not practical, make sure they understand the questions at the end of the lesson and have them do one of the "Dig Deeper" projects or another project.

You should be able to complete most of the 20 activities in a semester. Suppose you are on an 18-week time frame with science labs held once a week for two or three hours. Most investigations can be completed in an hour or less. Some of the shorter activities can be done on the same day or you may choose to do a teacher demonstration of a couple of the labs.

It is suggested that at least five hours a week be allotted to the investigations, contests, sharing of student projects, discussion of "What Have You Learned" questions, and research time. More time may be needed for some of the research and projects. Count projects, contests, and Dig Deeper activities equally. There are many possible activities from which students may choose.

Any time chemicals are used that might irritate eyes, safety glasses should be required. This is also a requirement for being around flames and other devices used for heating water or other chemicals. They are as important as safety belts are for children in a moving vehicle. Some activities should be done only as demonstrations led by an adult, but a student helper can assist if the student is wearing safety glasses and covering to protect clothes.

Refer students to textbooks or other references to help them answer questions, but also encourage them to think of their own explanations. It is not too early to help students understand that science is mostly about finding explanations for things they have observed and about finding patterns in nature. When controlled experiments are done, help them identify the controls and the variable.

Most of the supplies and equipment can be obtained locally. Also, supplies, videos of lessons, and other helps are available at www.investigatethepossibilities.org.

Objectives

1. The solar system is made up of the sun, planets, moons of planets, asteroid belt, TNOs, comets, and anything else held in orbit around the sun by its gravitation pull. The solar system is billions of miles across. It is measured in units called Astronomical Units (AU) which is equal to the distance from the earth to the sun.
2. The size of the solar system is too big to view it all at one time. We used a penny model to represent the distances of the planets to the sun, letting one penny equal one AU.

INVESTIGATION #1

What Is the Universe?

Think about This Michael and Alex alternated between catching fireflies in their back yard and trying to count the stars in the sky. Their parents had set up a refracting telescope to do some star watching, since the night was unusually clear. "Do you want to see something that looks like one star, but it's really many, many stars?" Mom asked.

"No way," Michael said doubtfully, as he hurried over to see. "It just looks like a funny-shaped fuzzy star to me."

"You're actually looking at about a billion stars that are incredibly far, far away from the earth," his mom continued. "They look like they are concentrated together. The truth is they are not close together at all, but just look like that way from earth."

Do you think the boys were seeing the very end of the universe? Do you think scientists have been able to locate the end of the universe by using powerful telescopes?

The Investigative Problems

How big is the solar system? How big is the universe? Can we make a model to help us understand these very large distances?

6

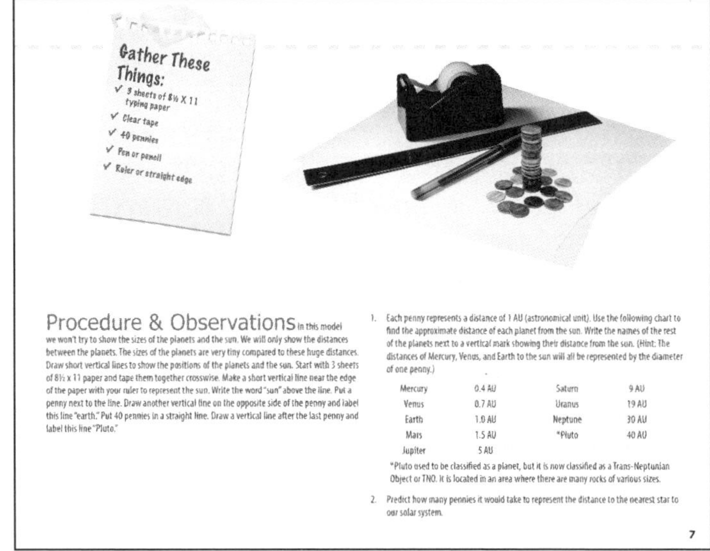

Procedure & Observations In this model we won't try to show the sizes of the planets and the sun. We will only show the distances between the planets. The sizes of the planets are very tiny compared to these huge distances. Draw short vertical lines to show the positions of the planets and the sun. Start with 3 sheets of 8½ x 11 paper and tape them together crosswise. Make a short vertical line near the edge of the paper with your ruler to represent the sun. Write the word "sun" above the line. Put a penny next to the line. Draw another vertical line on the opposite side of the penny and label this line "earth." Put 40 pennies in a straight line. Draw a vertical line after the last penny and label this line "Pluto."

1. Each penny represents a distance of 1 AU (astronomical unit). Use the following chart to find the approximate distance of each planet from the sun. Write the names of the rest of the planets next to a vertical mark showing their distance from the sun. (Hint: The distances of Mercury, Venus, and Earth to the sun will all be represented by the diameter of one penny.)

Mercury	0.4 AU	Saturn	9 AU
Venus	0.7 AU	Uranus	19 AU
Earth	1.0 AU	Neptune	30 AU
Mars	1.5 AU	*Pluto	40 AU
Jupiter	5 AU		

*Pluto used to be classified as a planet, but it is now classified as a Trans-Neptunian Object or TNO. It is located in an area where there are many rocks of various sizes.

2. Predict how many pennies it would take to represent the distance to the nearest star to our solar system.

7

3. Our solar system is part of the Milky Way Galaxy. There are at least 200 billion stars in the Milky Way Galaxy. Distances across the galaxy have to be measured in very large units, such as light years, which is the distance light travels in one year.
4. There are billions of galaxies in the universe. The universe seems to be millions of light years across.
5. Astronomers have not been able to identify the end of the universe and do not know how large it is. Evidence indicates the most distant stars may be millions of light years away.
6. Some of the methods used to estimate distances of objects in space are given.

Note

The following is an alternative way to model the size of the solar system. Find a large enough space that is safe for students to walk. If this activity is done, it should be supervised by a teacher or a parent, because it involves walking a long distance. If you only have a limited space, you can zigzag back and forth as long as the students understand they should be walking farther away from the "sun." Use something like a soccer ball to represent the sun. Use modeling clay to form balls to represent the planets. Mercury will only be a few millimeters in diameter. Jupiter will be a couple of centimeters in diameter. Use a reference book or the Internet to make the clay balls larger or smaller as needed. The exact sizes of the planets are not too important as long as they are relatively close. Place the clay models on index cards that are labeled with the names of the planets.

Begin by placing the "sun" (soccer ball) on the ground. Then take 7 steps away from the sun and place the clay ball representing Mercury on the ground on an index card labeled "Mercury." Take 6 more steps to Venus, placing the clay ball for Venus on an index card labeled "Venus." Repeat this process for each planet. Take 5 more steps to Earth; 10 more steps to Mars; 66 more steps to Jupiter; 78 more steps to Saturn; 173 more steps to Uranus; 195 more steps to Neptune; and 168 more steps to Pluto (which is now referred to as a "Trans-Neptunian Object" or TNO rather than a planet).

Be sure students realize that this is an imperfect model of the solar system. All of the planets are not on the same side of the sun at the same time. The model is only designed to help them visualize how far the planets are from the sun and to get an idea of the sizes of the planets.

Discussions

What are some of the things the sun, the moon, and the stars reveal to us about God?

The human body is made up of trillions of cells. Do you think a trillion stars in space are more awesome than the trillion cells that work together in your body in perfect harmony?

Read Isaiah 40:26. Do you find it amazing that God knows the name of every star?

The Science Stuff

The first four planets, Mercury, Venus, Earth, and Mars, are rocky planets. They are much closer to the sun than the four outer gas planets, Jupiter, Saturn, Uranus, and Neptune. Between Mars and Jupiter there is an asteroid belt composed of rocks of various sizes that are in orbit around the sun. Beyond Neptune, there is another belt of rocks, known as Trans-Neptunian Objects (or TNOs), which is where Pluto is found. Many of these rocky objects, like Pluto, have moons that orbit them.

You may be wondering why Pluto is no longer classified as a planet. Recently, two TNOs larger than Pluto were discovered, and it is likely that other larger TNOs will also be discovered in the future. So there is no reason to give special status to Pluto.

The penny model you made should help you better understand some things about the size of our solar system and how far apart the planets are. Like most models, it represents something too big or too complex to be seen, but it isn't perfect. For example, you probably know the planets are not in a row all on the same side of the sun.

Here are some important concepts the model can help you understand: The sun is 93,000,000 miles (150,000,000 km) away from Earth and is represented by one penny. This distance is also referred to as 1 astronomical unit (AU). Forty pennies represents about 4,000,000,000 miles (6,000,000,000 km) or 40 AU.

Incredibly, when we observe objects outside of our solar system, AU units are too small to work well. For these extremely large distances, astronomers use a different unit called a light-year. This is the distance that light travels in one year. One light year is equal to about 6,250,000,000,000 miles (10,000,000,000,000 km).

For comparison, think about these examples. Light can travel about 186,000 miles (300,000 km) in one second. It takes about 8 minutes for light to travel from the sun to the earth. It takes an average of about 5.5 hours for light to travel from the sun to Pluto. Look at the penny model again to see the distances that are represented.

The next nearest star to our sun is actually a star system made up of a pair (possibly triple set) of stars, known as the Alpha Centauri system. These stars are more than 4 light years away. Using our penny model, we would need a line of pennies a little over 3 miles (4.8 km) long to show where this pair of stars is.

Our sun and the Alpha Centauri pair are only 3 of the stars in our Milky Way Galaxy. Some astronomers have estimated that there may be more than 300 billion stars in our Milky Way Galaxy. We would have to use the entire Pacific Ocean to extend the penny model and make an accurate model of the whole galaxy!

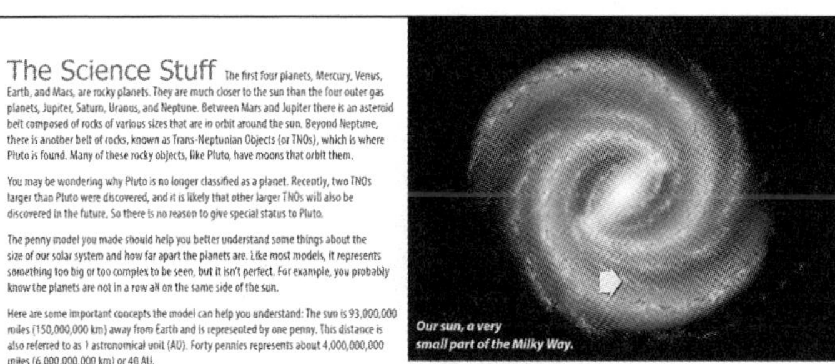

Our sun, a very small part of the Milky Way.

Galaxies are organized into clusters of galaxies. Our Milky Way Galaxy is part of a group of 30 or more galaxies that are bound together by mutual gravitational attraction. This group is known as the Local Group.

The word *universe* is a term that includes all the galaxies in space. There seems to be from 100 to 200 billion other galaxies in the universe, but the number keeps changing as telescopes and technology make it possible to detect more stars and galaxies. Some of the points of light we can see in the night sky are really far away galaxies made up of billions of individual stars. The universe is so big that it is impossible for us to even imagine its size.

It is easy to be confused about all of this information, so let's review the main points. Starting with our solar system, you should understand that there are 8 planets (9 if you count Pluto) that travel around the sun. Most of the planets have their own moons. The planets, their moons, comets, asteroids, TNOs, dust, and everything else that is captured by the sun's gravitational force are part of our solar system.

Our solar system is part of the Milky Way Galaxy, which is part of a cluster of galaxies known as the Local Group. Our Milky Way Galaxy is only one of the hundreds of billions of galaxies in the whole universe.

8

Making Connections

Astronomers have calculated that there are about 6,000 stars that can be seen under the best night time conditions, but the actual number is much greater than that. Scientists have always been fascinated with the number of stars in the universe, and there have been many attempts to count them by astronomers. The number is unbelievably huge. For example in our Milky Way Galaxy, it is believed that there may be 200 billion stars.

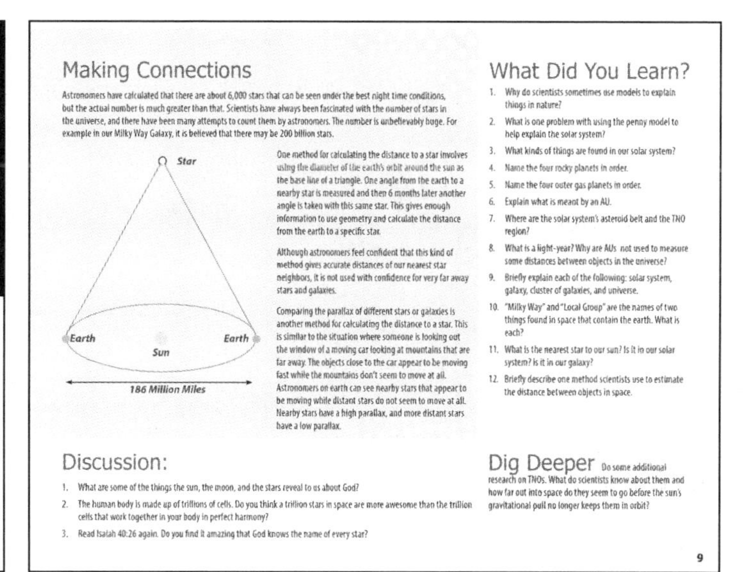

One method for calculating the distance to a star involves using the diameter of the earth's orbit around the sun as the base line of a triangle. One angle from the earth to a nearby star is measured and then 6 months later another angle is taken with this same star. This gives enough information to use geometry and calculate the distance from the earth to a specific star.

Although astronomers feel confident that this kind of method gives accurate distances of our nearest star neighbors, it is not used with confidence for very far away stars and galaxies.

Comparing the parallax of different stars or galaxies is another method for calculating the distance to a star. This is similar to the situation where someone is looking out the window of a moving car looking at mountains that are far away. The objects close to the car appear to be moving fast while the mountains don't seem to move at all. Astronomers on earth can see nearby stars that appear to be moving while distant stars do not seem to move at all. Nearby stars have a high parallax, and more distant stars have a low parallax.

Discussion:

1. What are some of the things the sun, the moon, and the stars reveal to us about God?
2. The human body is made up of trillions of cells. Do you think a trillion stars in space are more awesome than the trillion cells that work together in your body in perfect harmony?
3. Read Isaiah 40:26 again. Do you find it amazing that God knows the name of every star?

What Did You Learn?

1. Why do scientists sometimes use models to explain things in nature?
2. What is one problem with using the penny model to help explain the solar system?
3. What kinds of things are found in our solar system?
4. Name the four rocky planets in order.
5. Name the four outer gas planets in order.
6. Explain what is meant by an AU.
7. Where are the solar system's asteroid belt and the TNO region?
8. What is a light-year? Why are AUs not used to measure some distances between objects in the universe?
9. Briefly explain each of the following: solar system, galaxy, cluster of galaxies, and universe.
10. "Milky Way" and "Local Group" are the names of two things found in space that contain the earth. What is each?
11. What is the nearest star to our sun? Is it in our solar system? Is it in our galaxy?
12. Briefly describe one method scientists use to estimate the distance between objects in space.

Dig Deeper

Do some additional research on TNOs. What do scientists know about them and how far out into space do they seem to go before the sun's gravitational pull no longer keeps them in orbit?

9

What Did You Learn?

1. Why do scientists sometimes use models to explain things in nature? *Models help students to visualize things that are too large or too complex to see.*
2. What is one problem with using the penny model to help explain the solar system? *The planets are not all lined up on the same side of the sun.*
3. What kind of things are found in our solar system? *The planets, their moons, comets, asteroids, TNOs, dust, and everything else that is captured by the sun's gravitational force.*
4. Name the four rocky planets in order. *Mercury, Venus, Earth, and Mars*
5. Name the four outer gas planets in order. *Jupiter, Saturn, Uranus, and Neptune*
6. Explain what is meant by an AU. *AU stands for Astronomical Unit. This is the distance from the sun to the earth and is equal to 93,000,000 miles or 150,000,000 km.*
7. Where are the solar system's asteroid belt and the TNO region? *The asteroid belt is located between Mars and Jupiter. The TNO region is beyond Neptune.*
8. What is a light-year? Why are AUs not used to measure some distances between objects in the universe? *A light-year is the distance light can travel through space in a year. This distance is equal to about 6 trillion miles or 10 trillion km. AU units are too small to be convenient for measuring very large distances.*
9. Briefly explain each of the following: solar system, galaxy, cluster of galaxies, and universe. *The solar system is made up of the sun and the planets that orbit the sun, along with everything else held in place by the sun's gravity. A galaxy is made up of millions or billions of stars that are bound together by gravitational attraction. A cluster of galaxies is a group of galaxies that are relatively close together. The universe is a term used to include everything that exists in the natural world.*
10. "Milky Way" and "Local Group" are the names of two things found in space that contain the earth. What is each? *The Milky Way is the name given to the galaxy that contains our solar system. Local Group is the name of a cluster of a few dozen galaxies that contains our Milky Way Galaxy.*
11. What is the nearest star to our sun? Is it in our solar system? Is it in our galaxy? *The nearest star to our sun is Alpha Centauri (probably a pair or triplet of stars). It is not in our solar system, but it is in our galaxy.*
12. Briefly describe one method scientists use to estimate the distance between objects in space. *Two methods are described in Making Connections. Students should choose one of these to describe: measuring angles between earth and a star, and the parallax method.*

Investigation #2

Spreading Out the Heavens

Think about This Julie and Justin were laughing hysterically as their father showed an old video of them playing on a slide. He showed the clip in forward motion and then showed it in reverse. "Do it again," they said in unison. Their big sister Anita laughed along with them. Then she thought how neat it would be if a video of their lives for the past year could be shown in reverse to show how much everyone had grown and changed over the year.

Her father said, "Let's take this idea a little further. What if there was a video of the entire universe that could be shown in reverse right back to the beginning of everything? What do you think we'd see at the very beginning?"

Anita remembered seeing a TV program once about the big-bang theory. It showed the beginning of the universe as a small dense ball. Then, about 15 billion years ago, the ball exploded and everything began to move out in all directions forming stars, planets, moons, and all the other objects in the universe. Do you think, if it were possible to see a movie of the universe in reverse, it would begin as a small dense "ball" 15 billion years ago?

The Investigative Problems

Is the universe expanding? What are some possible explanations for an expanding universe?

10

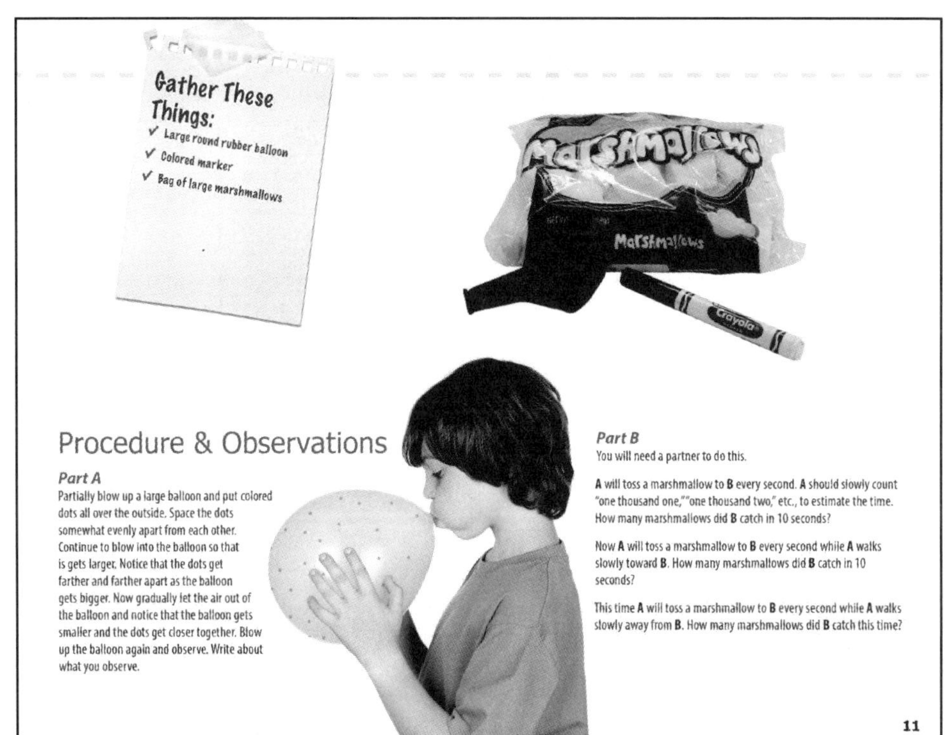

Gather These Things:

✓ Large round rubber balloon
✓ Colored marker
✓ Bag of large marshmallows

Procedure & Observations

Part A

Partially blow up a large balloon and put colored dots all over the outside. Space the dots somewhat evenly apart from each other. Continue to blow into the balloon so that is gets larger. Notice that the dots get farther and farther apart as the balloon gets bigger. Now gradually let the air out of the balloon and notice that the balloon gets smaller and the dots get closer together. Blow up the balloon again and observe. Write about what you observe.

Part B

You will need a partner to do this.

A will toss a marshmallow to **B** every second. **A** should slowly count "one thousand one,""one thousand two," etc., to estimate the time. How many marshmallows did **B** catch in 10 seconds?

Now **A** will toss a marshmallow to **B** every second while **A** walks slowly toward **B**. How many marshmallows did **B** catch in 10 seconds?

This time **A** will toss a marshmallow to **B** every second while **A** walks slowly away from **B**. How many marshmallows did **B** catch this time?

11

Objectives

1. Astronomers began to recognize a sun-centered solar system in the 1500s as they studied the movements of planets. Telescopes had been invented and were used by Copernicus, Galileo, and Kepler during this time.
2. In the early 1900s, Hubble discovered other galaxies and noted that all their spectra showed a red shift. This was a clue that the galaxies were expanding throughout the universe.
3. The big-bang theory was proposed based on the red shifts in the light spectra and a 2.7° K temperature throughout space. The nebula hypothesis was proposed to explain how the stars and the planets formed after the big bang. According to these proposals, billions of years would be required.
4. Creation scientists have proposed alternative explanations that only require thousands of years. They also accept the idea that the stars were spread out as they were created.

Notes

History books tend to portray the disagreements between Galileo and the Catholic pope as an ongoing battle between science and religion. This is greatly exaggerated in many accounts of history. Try to help students see that true science and the Bible are not in conflict. Conflicts arise from how evidence is interpreted. Darwinian evolution and billions of years of slow changes are the two main areas where science and the Bible have different interpretations of origins.

What Did You Learn?

1. What scientist discovered that there were other galaxies in the universe in addition to the galaxy our earth is in? *Edwin Hubble*
2. What evidence did Edwin Hubble discover that caused him to conclude that galaxies are moving and getting farther away from the earth? *He collected spectra of light from 46 galaxies and noted that there was always a red shift in the colors of the visible spectrum that came from these galaxies.*

The Science Stuff

There has been an assortment of ideas through the ages about how to explain the universe, but there are still more questions than answers. Some of the first real clues about how the solar system works came from scientists like Copernicus (1473–1543), Galileo (1564–1642), and Kepler (1571–1630). At the time when Copernicus cautiously proposed that the planets moved around the sun, he did not have the benefit of telescopes. Both Galileo and Kepler were able to study the planets with telescopes. The explanation for movements of planets and stars caused people to eventually shift from believing in an earth-centered universe to accepting a sun-centered solar system. The change to a sun-centered solar system took a long time, but by the time of Isaac Newton (1642–1727), the evidence had persuaded most scientists it was true.

Another major shift in viewing the universe came as Edwin Hubble (1889–1953) made use of more powerful telescopes and other kinds of technology during the early 1900s. At first Hubble agreed with other astronomers of his day that the solar system, the stars, comets, asteroids, and nebulae were all part of the same galaxy. At this time, it was thought that there was nothing outside of our galaxy.

But in 1923, as Hubble began studying a fuzzy patch of sky called the Andromeda Nebula, he found that it contained individual stars. After making many observations and then doing some mathematical calculations, he finally concluded that he was viewing a set of stars that made up another galaxy completely separate from the one the earth was in. In the next few years, Hubble was able to identify several other galaxies. By 1929, most astronomers had come to believe that our Milky Way Galaxy was only one of millions of galaxies in the universe.

Hubble discovered that not all galaxies are alike. He found elliptical galaxies, spiral galaxies, and barred spiral galaxies. We know today that there are even more different kinds of galaxies than Hubble imagined.

Eventually, Hubble collected spectra of light from 46 galaxies. He noted that there was always a red shift in the colors of the visible spectrum that came from these galaxies.

Since red is the longest visible wavelength, a red shift in the spectrum would indicate that the light from these galaxies was being stretched out as it reached the earth. This information led scientists to believe that all of the galaxies were moving away from the earth and getting farther apart over time. If the spectrum colors had always shifted toward blue/violet, the shortest wavelengths, this would have been a clue that the galaxies were moving toward the earth and getting closer together. If the light the galaxies emit was stretched out by the time it reached the earth, the light spectrum would show a red shift. This is a clue that the universe is expanding.

The activity with the marshmallows should help you understand this concept. The light (or sound) waves are squeezed together as one object approaches another. The light (or sound) waves are spread out as one object moves away from another one. The marshmallows were easier to catch as the pitcher moved away from the catcher, representing how the waves were stretched out a little.

Not long after the red shift discovery, the "big-bang" theory was introduced. Many astronomers reasoned that if galaxies in the universe are now expanding, there must have been a time in the past when they were closer together. They kept trying to rewind time and concluded that the galaxies must have started out from the same place. They reasoned that their expanding motion must have started from an explosion. Hence, the name the "big bang."

Soon, another theory known as the nebula theory was made based on the big-bang theory. According to this theory, after the big bang occurred, large clouds of dust particles were thrown into space and began to spin. Within these spinning clouds, other whirlpools formed. Then the particles in these whirlpools condensed to form galaxies, stars, planets, and other objects in space.

There are problems with the big-bang theory, the nebula hypothesis, and billions of years to create everything. However, all three principles continue to be accepted by mainstream scientists.

An alternative explanation that is consistent with Scripture is being studied by some astronomers. They are looking at the possibility that the originally created universe was much smaller than it is today. And, at some point after creation, it expanded to the present size, causing light rays to also be "stretched" in the process. Some creation scientists believe this process might be something like how you observed the spots on the balloon expand, as the balloon got bigger.

Hubble Spacecraft

Elliptical galaxies

Spiral galaxies

Barred spiral galaxies

12

Johannes Kepler 1571-1630

Nicolaus Copernicus 1473-1543

Galileo Galilei 1564-1642

Sir Isaac Newton 1642-1747

Edwin Hubble 1889–1953

Making Connections

Recall that *parallax* is a method for estimating extremely distant objects in space and that such objects may appear not to be moving. When objects are trillions of miles away, they appear to be staying still even though they are probably moving.

The idea that the universe was stretched out is found in several places in the Bible. Isaiah wrote that God stretched out the heavens like a curtain. Notice that this idea is like the big bang in suggesting that there was once an expanding of the sky.

The appearance of a red shift in the light reaching the earth from distant galaxies is based on an assumption that light from these galaxies is moving away from the earth. This should work in the same way that radar uses the Doppler effect to give readings about the speed and direction of moving clouds. Waves from the clouds are compressed as they move toward the radar and are stretched out as they move away from the radar.

Dig Deeper

What are the differences between astronomers and astrologers? Try to find the names of early civilizations that studied the heavenly bodies. (Keep in mind that some of the early survivors of the Flood were expert astronomers.)

Find out how a Doppler radar works and how weather forecasters use them in predicting weather. Compare Doppler weather radar with instruments that astronomers use to measure shifts in light spectra.

Position yourself on a sidewalk. Have a parent drive a car down the road next to the sidewalk while continuously blowing the car horn. Describe how the pitch of the car horn changes as the car approaches and then moves away. The change in pitch is known as the Doppler effect.

Draw shapes of different kinds of galaxies.

What Did You Learn?

1. What scientist discovered that there were other galaxies in the universe in addition to the galaxy our earth is in?
2. What evidence did Edwin Hubble discover that caused him to conclude that galaxies are moving and getting farther away from the earth?
3. Before the time of Hubble, did scientists believe all the stars in the universe were in the same galaxy?
4. Which color in the visible spectrum has the longest wavelength?
5. Is the bluish/violet end of the visible spectrum made up of shorter waves or longer waves?
6. What major shift in thinking about the solar system came from scientists like Copernicus, Galileo, and Kepler?
7. The "big-bang" theory is based on what main piece of evidence? Does this prove that the big bang actually happened?
8. Briefly tell about the "nebula theory." Does it attempt to explain the origin of all the stars, planets, moons, comets, rocks, and dust in the universe?
9. What instrument was available for Galileo, Kepler, and Hubble to use that Copernicus did not have?
10. Give the shape of two different kinds of galaxies.
11. All galaxies appear to be moving. Why are we unable to look at them and tell that they are moving?

Pause and Think

There are numerous references in Job, Isaiah, Jeremiah, and Zechariah that refer to a mighty act of God in which He spread out the stars in the sky.

Isaiah 48:13 Isaiah 51:13 Isaiah 40:22 Job 26:7
Isaiah 42:5 Isaiah 44:24 Isaiah 45:12 Job 37:18
Jeremiah 10:12 Zechariah 12:1

13

3. Before the time of Hubble, did scientists believe all the stars in the universe were in the same galaxy? *Yes*
4. Which color in the visible spectrum has the longest wavelength? *Red*
5. Is the bluish/violet end of the visible spectrum made up of shorter waves or longer waves? *Shorter*
6. What major shift in thinking about the solar system came from scientists like Copernicus, Galileo, and Kepler? *They concluded that the sun was the center of the solar system, and that the earth and the planets revolved around the sun.*
7. The "big-bang" theory is based on what main piece of evidence? Does this prove that the big bang actually happened? *A red shift in the light spectra of galaxies is the main piece of evidence for the big-bang theory. This does not prove the big bang actually happened. There are other logical explanations for this evidence.*
8. Briefly tell about the "nebula theory." Does it attempt to explain the origin of all the stars, planets, moons, comets, rocks, and dust in the universe? *The nebula theory proposes that, following the big bang, there were swirling clouds throughout space. Eventually these clouds formed swirling eddies of gas that condensed into stars and planets and all the other objects in space.*
9. What instrument was available for Galileo, Kepler, and Hubble to use that Copernicus did not have? *Telescopes*
10. Give the shape of two different kinds of galaxies. *Shapes include spiral galaxies, elliptical galaxies, barred spiral galaxies, ring galaxies, and other shapes.*
11. All galaxies appear to be moving. Why are we unable to look at them and tell that they are moving? *The parallax effect makes it difficult to detect movement in objects that are extremely far away.*

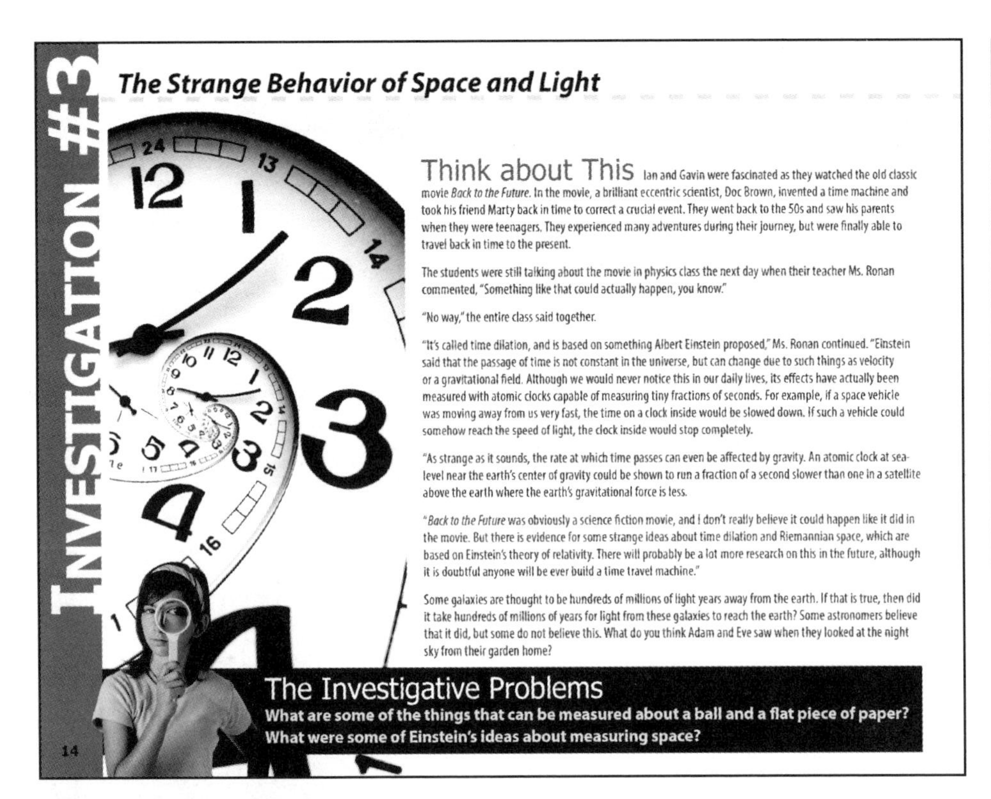

Investigation #3

The Strange Behavior of Space and Light

Think about This Ian and Gavin were fascinated as they watched the old classic movie *Back to the Future.* In the movie, a brilliant eccentric scientist, Doc Brown, invented a time machine and took his friend Marty back in time to correct a crucial event. They went back to the 50s and saw his parents when they were teenagers. They experienced many adventures during their journey, but were finally able to travel back in time to the present.

The students were still talking about the movie in physics class the next day when their teacher Ms. Ronan commented, "Something like that could actually happen, you know."

"No way," the entire class said together.

"It's called time dilation, and is based on something Albert Einstein proposed," Ms. Ronan continued. "Einstein said that the passage of time is not constant in the universe, but can change due to such things as velocity or a gravitational field. Although we would never notice this in our daily lives, its effects have actually been measured with atomic clocks capable of measuring tiny fractions of seconds. For example, if a space vehicle was moving away from us very fast, the time on a clock inside would be slowed down. If such a vehicle could somehow reach the speed of light, the clock inside would stop completely.

"As strange as it sounds, the rate at which time passes can even be affected by gravity. An atomic clock at sea-level near the earth's center of gravity could be shown to run a fraction of a second slower than one in a satellite above the earth where the earth's gravitational force is less.

"*Back to the Future* was obviously a science fiction movie, and I don't really believe it could happen like it did in the movie. But there is evidence for some strange ideas about time dilation and Riemannian space, which are based on Einstein's theory of relativity. There will probably be a lot more research on this in the future, although it is doubtful anyone will be ever build a time travel machine."

Some galaxies are thought to be hundreds of millions of light years away from the earth. If that is true, then did it take hundreds of millions of years for light from these galaxies to reach the earth? Some astronomers believe that it did, but some do not believe this. What do you think Adam and Eve saw when they looked at the night sky from their garden home?

The Investigative Problems

What are some of the things that can be measured about a ball and a flat piece of paper?
What were some of Einstein's ideas about measuring space?

14

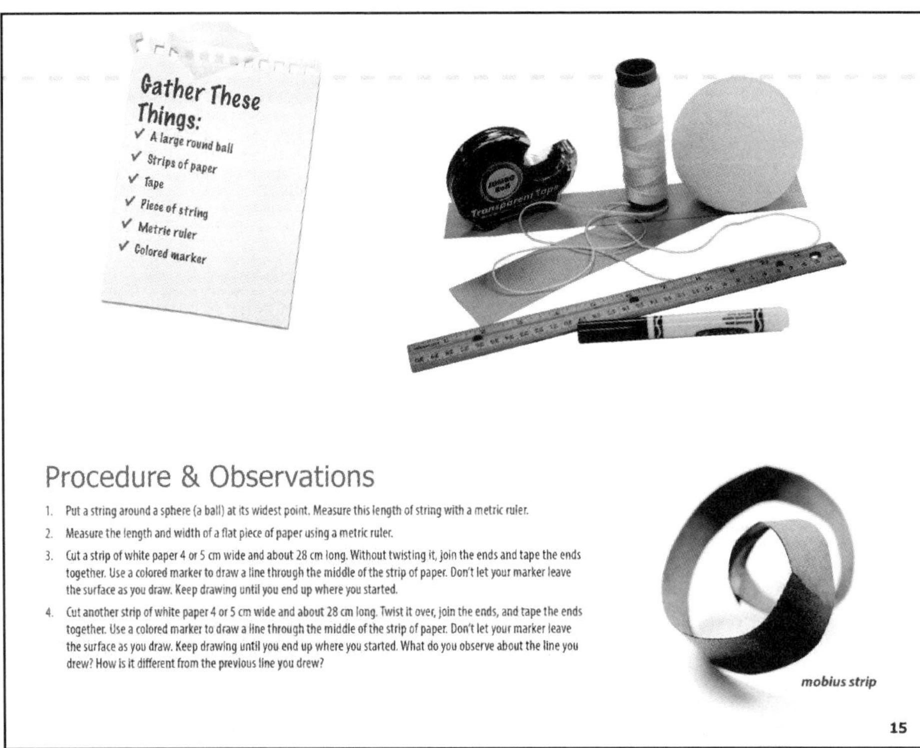

Gather These Things:
✓ A large round ball
✓ Strips of paper
✓ Tape
✓ Piece of string
✓ Metric ruler
✓ Colored marker

Procedure & Observations

1. Put a string around a sphere (a ball) at its widest point. Measure this length of string with a metric ruler.
2. Measure the length and width of a flat piece of paper using a metric ruler.
3. Cut a strip of white paper 4 or 5 cm wide and about 28 cm long. Without twisting it, join the ends and tape the ends together. Use a colored marker to draw a line through the middle of the strip of paper. Don't let your marker leave the surface as you draw. Keep drawing until you end up where you started.
4. Cut another strip of white paper 4 or 5 cm wide and about 28 cm long. Twist it over, join the ends, and tape the ends together. Use a colored marker to draw a line through the middle of the strip of paper. Don't let your marker leave the surface as you draw. Keep drawing until you end up where you started. What do you observe about the line you drew? How is it different from the previous line you drew?

mobius strip

15

Objectives

1. Newton's laws of motion and gravitation explain the motion of objects when they are on an object that is assumed to be stationary, not moving, and somewhat flat.
2. Einstein didn't change Newton's laws, but he expanded them. He formulated ideas about the unexpected behavior of objects traveling near the speed of light. He predicted certain behaviors when gravitation becomes very strong.
3. Some ideas are briefly discussed about how light traveling from millions of light years away could have reached the earth in short periods of time.

Note

Some of the ideas in this lesson may give us a little scientific insight into how stars originally formed, but the creation of light on day 1 and stars on day 4 are basically supernatural events. Remind students that there are no scientific explanations for how Jesus walked on water, how He instantly healed many sick people, how He rose from the dead, and how He rose up in a cloud as He returned to heaven. These things not only show that He created the natural laws and processes, but that He also has authority over them.

What Did You Learn?

1. Why is a Mobius strip difficult to measure? *It only has one surface, since no matter where you start, a marker will make a continuous line that is found on both sides of the paper.*
2. According to Einstein's special theory of relativity, what is the fastest speed that anything can reach? *The speed of light in a vacuum*

The Science Stuff

There are specific things that can be measured for a flat sheet of paper, and a ball. Lengths and surface area are two of the things that can be calculated from what you measured.

The twisted strip is known as a Mobius strip. Its boundaries are not as easy to describe or measure as the other objects, because it only has one surface! No matter where you start, if you don't lift your marker and you stay in the middle of the strip, a continuous surface is found on both sides of the paper.

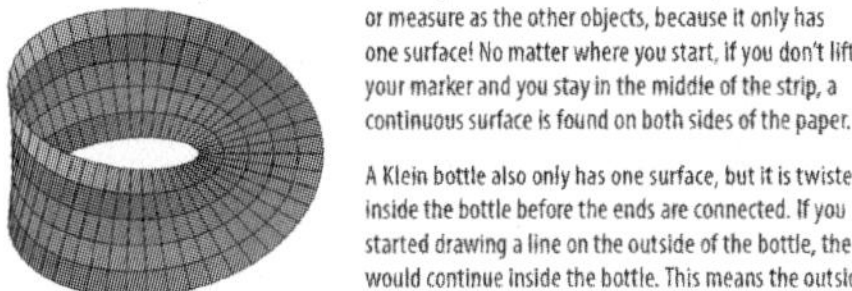

Mobius strip

A Klein bottle also only has one surface, but it is twisted inside the bottle before the ends are connected. If you started drawing a line on the outside of the bottle, the line would continue inside the bottle. This means the outside of the bottle becomes the inside!

The Klein bottle illustrates something known as Riemannian geometry. It is too complex to describe here, but astronomers have become quite interested in it as a possible way to explain space and time. Many astronomers believe that the whole universe must be curved. They believe the speed of an object, distance, time, gravity, and the speed of light are all interrelated, so that one thing can affect another.

Klein bottle

Most of what we know about forces and motion are based on Newton's laws. They are still dependable throughout the universe as long as they are viewed from a location that is assumed to be flat and not moving. Einstein found a new way of expanding Newton's laws of motion by taking into account the fact that everything in the universe is moving and velocity and gravity can affect time.

Einstein proposed that we should not assume that light coming from space always travels in a straight line. He showed that some of the things we measure are different in different situations, and even the rate at which time passes is not fixed.

Einstein proposed his special theory of relativity in 1905 and his general theory of relativity in 1915. His first theory showed that Newton's three laws of motion don't always work when objects approach the speed of light. His second theory showed that Newton's law of universal gravitation doesn't always work when the gravitational force becomes very strong.

Christian astronomers who doubt that the universe is billions of years old are cautiously considering some of Einstein's ideas about the relationships between light, space, time, mass, and gravity. Some of these ideas are very complex, so we will only briefly mention a few of them.

These ideas are not intended to provide a natural explanation for a supernatural act. We may never understand how God created such an immense fully formed universe, so that the first people on earth saw a sky full of stars. Nevertheless, we believe they are worth mentioning.

(1) Some creation models include the idea of time dilation so that things in space moving near the speed of light might age much faster than the actual time on earth shows.

(2) Space could be explained in terms of Riemannian geometry where space is curved and never-ending. Some calculations have shown that if this is true, light could reach earth very quickly from any part of the universe.

(3) Another idea is that in an early stage of creation, light may have traveled more rapidly than it does now.

(4) Light was created on the first day of creation before the sun and the stars were formed. Light would have filled the universe just like fully mature trees and animals were created on day 6 and filled the earth.

(5) Some astronomers have suggested that the earth is near the center of the universe in a "gravitational well." Gravitation is one of the things that affect the passage of time. If the earth is in such a "well," then light that would take billions of years to reach earth (according to clocks in deep space) might reach earth in only thousands of years (according to clocks on earth), or even instantaneously!

(6) Another possibility suggests that both the stars and the light they gave off were stretched out across the universe after, or as, they were created.

(7) One, or a combination, of these explanations may be correct. However, the true explanation may be something that cannot be explained on the basis of the natural processes that operate today. Science can only investigate the way in which the universe operates today. It has no way of explaining supernatural events. On the other hand, science has no right to declare that supernatural events never happened!

16

The general theory of relativity concerns gravitation and was discussed in a publication by Albert Einstein in 1916.

Here is an artist concept of gravity showing a probe orbiting the Earth to measure spacetime: a four-dimensional idea of the universe that includes height, width, length, and time.

Making Connections

Based on Genesis 1:1, creation scientists believe the heavens and the earth had a beginning. Based on the big-bang theory, most evolutionary scientists also believe that the universe had a beginning. Creation and evolutionary scientists disagree about the length of time this took and how a random explosion could result in orderly balanced systems, such as the solar system.

Dig Deeper

According to Einstein's special theory of relativity, there is nothing that can travel faster than the speed of light in a vacuum. Do some research on muons. Explain why scientists are puzzled by their behavior.

Read about Einstein's theories of relativity in a book written for students. Explain some of the main ideas of these theories. (Keep it simple.)

Pause and Think

We should remember that the creation of light on day 1 and the creation of the stars on day 4 may have left some physical evidence, but these events are basically supernatural events. The Bible tells of many other supernatural events that Jesus caused to happen. For example, He walked on water, He instantly healed sick people, He rose from the dead, and He ascended to heaven on a cloud. These things not only show that He is the Creator of natural laws and processes, but that He also has authority over them.

What Did You Learn?

1. Why is a Mobius strip difficult to measure?
2. According to Einstein's special theory of relativity, what is the fastest speed that anything can reach?
3. There were six hypothetical explanations for how light was able to reach across the universe in less than millions of years. Briefly explain one of these explanations.
4. What is one way in which Einstein's theories of relativity add to Newton's laws of motion?
5. Are Newton's laws of motion always correct? Under what conditons might they not be correct?
6. Does Einstein believe that light always travels in a straight line when it passes through outer space?
7. Einstein studied time and what four other factors as part of his theories of relativity?
8. What did Einstein name his two theories of relativity?
9. What is the basic difference between Einstein's two theories of relativity?

17

3. There were six hypothetical explanations for how light was able to reach across the universe in less than millions of years. Briefly explain one of these explanations. *Refer to the chapter for this answer.*

4. What is one way in which Einstein's theories of relatively add to Newton's laws of motion? *He expanded Newton's laws of motion by taking into account the fact that everything in the universe is moving. He found that velocity and gravity can affect time. He showed that some of the things we measure are different in different situations.*

5. Are Newton's laws of motion always correct? Under what conditions might they not be correct? *They are still dependable throughout the universe as long as they are viewed from a location that is assumed to be flat and not moving. Solutions may change when the fact that everything in the universe is moving is taken into account. Solutions may also change when the fact that velocity and gravity can affect time is taken into account.*

6. Does Einstein believe that light always travels in a straight line when it passes through outer space? *He said we should not assume that light coming from space always travels in a straight line.*

7. Einstein studied time and what four other factors as part of his theories of relativity? *Light, space, mass, and gravity*

8. What did Einstein name his two theories of relativity? *Special theory of relativity and general theory of relativity*

9. What is the basic difference between Einstein's two theories of relativity? *His first theory showed that Newton's three laws of motion don't always work when objects approach the speed of light. His second theory showed that Newton's law of gravitation doesn't always work when the gravitational force becomes very strong.*

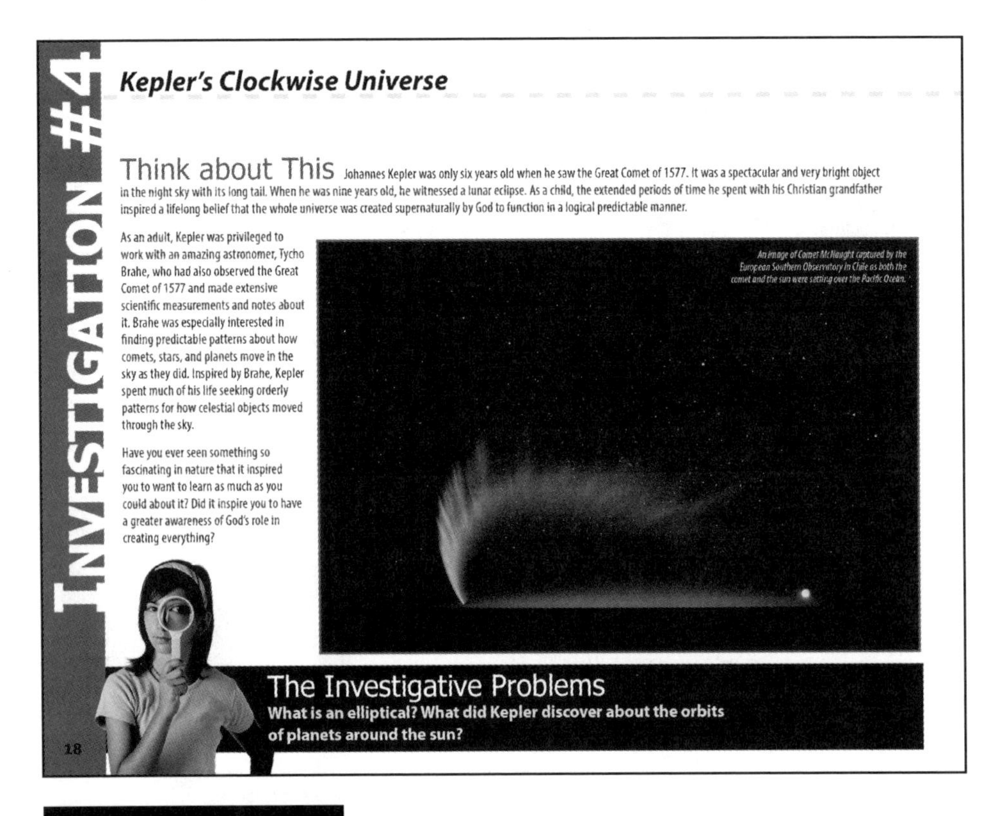

Investigation #4

Kepler's Clockwise Universe

Think about This

Johannes Kepler was only six years old when he saw the Great Comet of 1577. It was a spectacular and very bright object in the night sky with its long tail. When he was nine years old, he witnessed a lunar eclipse. As a child, the extended periods of time he spent with his Christian grandfather inspired a lifelong belief that the whole universe was created supernaturally by God to function in a logical predictable manner.

As an adult, Kepler was privileged to work with an amazing astronomer, Tycho Brahe, who had also observed the Great Comet of 1577 and made extensive scientific measurements and notes about it. Brahe was especially interested in finding predictable patterns about how comets, stars, and planets move in the sky as they did. Inspired by Brahe, Kepler spent much of his life seeking orderly patterns for how celestial objects moved through the sky.

Have you ever seen something so fascinating in nature that it inspired you to want to learn as much as you could about it? Did it inspire you to have a greater awareness of God's role in creating everything?

An image of Comet McNaught captured by the European Southern Observatory in Chile as both the comet and the sun were setting over the Pacific Ocean.

The Investigative Problems

What is an elliptical? What did Kepler discover about the orbits of planets around the sun?

18

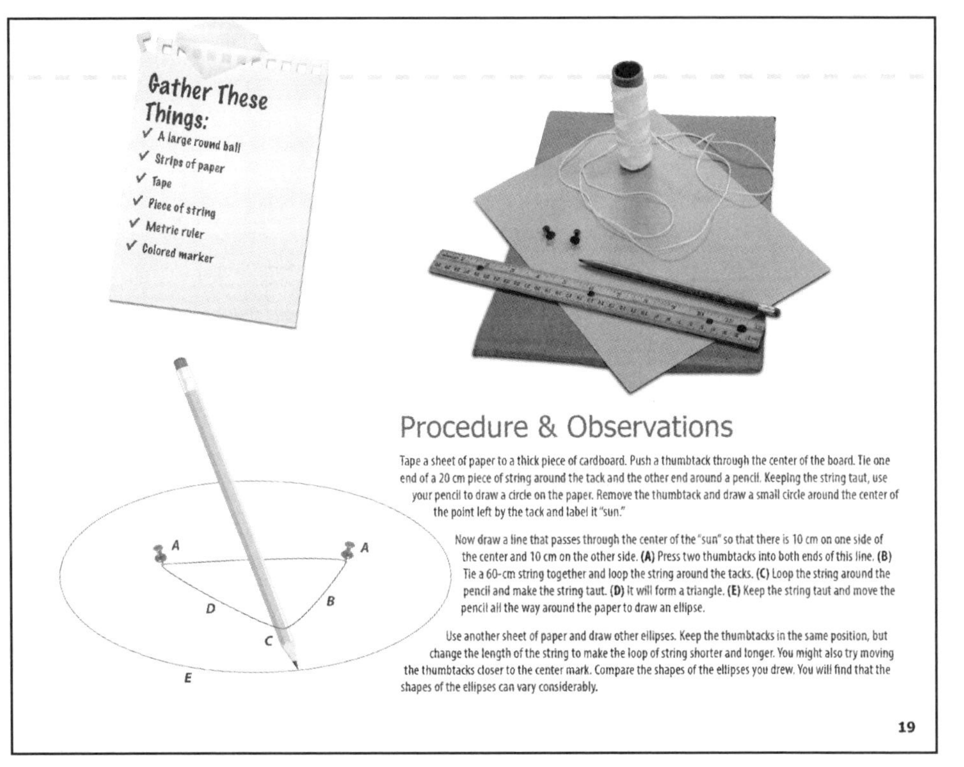

Procedure & Observations

Tape a sheet of paper to a thick piece of cardboard. Push a thumbtack through the center of the board. Tie one end of a 20 cm piece of string around the tack and the other end around a pencil. Keeping the string taut, use your pencil to draw a circle on the paper. Remove the thumbtack and draw a small circle around the center of the point left by the tack and label it "sun."

Now draw a line that passes through the center of the "sun" so that there is 10 cm on one side of the center and 10 cm on the other side. **(A)** Press two thumbtacks into both ends of this line. **(B)** Tie a 60-cm string together and loop the string around the tacks. **(C)** Loop the string around the pencil and make the string taut. **(D)** It will form a triangle. **(E)** Keep the string taut and move the pencil all the way around the paper to draw an ellipse.

Use another sheet of paper and draw other ellipses. Keep the thumbtacks in the same position, but change the length of the string to make the loop of string shorter and longer. You might also try moving the thumbtacks closer to the center mark. Compare the shapes of the ellipses you drew. You will find that the shapes of the ellipses can vary considerably.

19

Objectives

1. Copernicus challenged the current belief that the sun, the stars, and the planets orbited the earth. He proposed that the planets orbited the sun instead.
2. Kepler formulate three basic laws about how planets move in space. He determined that the planets move in elliptical orbits rather than in round circles.

Note

Time lines can be invaluable to students, but usually require teachers to help them draw lessons from them. Note that many discoveries about astronomy were made in a relatively short period of time, as one discovery built on earlier ones. Einstein's ideas came much later, but they were still dependent upon these earlier scientists. When a group of great scientists, who studied the same topics, lived in the same time period, they usually learned from each other. For some reason, fellow astronomers did not consider Kepler's ideas very important at first. Later, his ideas became foundational for astronomy and are still used today.

What Did You Learn?

1. Before the time of Copernicus, scientists believed that the sun, the stars, the moon, and the planets orbited the earth. Explain what Copernicus proposed that disagreed with this. *He proposed that the earth and the planets orbited the sun.*
2. Copernicus believed that the planets orbited the sun in perfect circles. What kind of orbit did Kepler propose that planets followed? *Kepler proposed that the planets followed an elliptical orbit around the sun.*
3. What is an elliptical orbit? *An elliptical orbit contains two focus points, but neither of the focal points is in its center. The sun is located at one of the focal points for the planets that follow an elliptical orbit around the sun.*
4. The earth gets a little closer to the sun at one phase of its orbit. Does the earth get warmer when it is closer to the sun? *No*

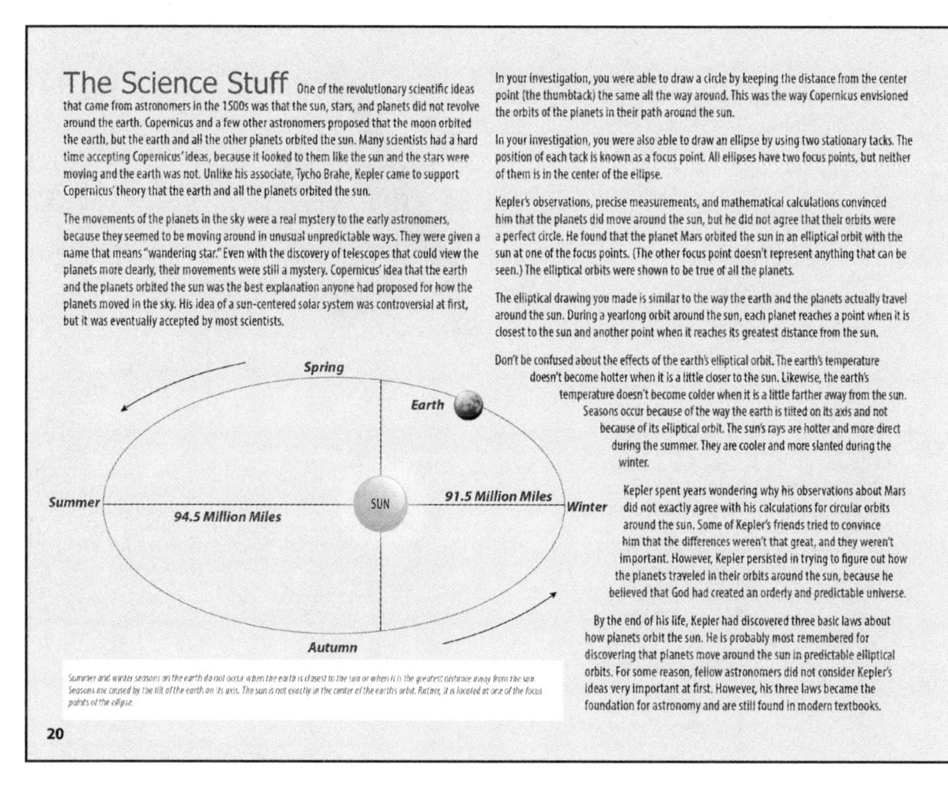

The Science Stuff

One of the revolutionary scientific ideas that came from astronomers in the 1500s was that the sun, stars, and planets did not revolve around the earth. Copernicus and a few other astronomers proposed that the moon orbited the earth, but the earth and all the other planets orbited the sun. Many scientists had a hard time accepting Copernicus' ideas, because it looked to them like the sun and the stars were moving and the earth was not. Unlike his associate, Tycho Brahe, Kepler came to support Copernicus' theory that the earth and all the planets orbited the sun.

The movements of the planets in the sky were a real mystery to the early astronomers, because they seemed to be moving around in unusual unpredictable ways. They were given a name that means "wandering star." Even with the discovery of telescopes that could view the planets more clearly, their movements were still a mystery. Copernicus' idea that the earth and the planets orbited the sun was the best explanation anyone had proposed for how the planets moved in the sky. His idea of a sun-centered solar system was controversial at first, but it was eventually accepted by most scientists.

Summer and winter seasons on the earth do not occur when the earth is closest to the sun or when it is the greatest distance away from the sun. Seasons are caused by the tilt of the earth on its axis. The sun is not exactly in the center of the earth's orbit. Rather, it is located at one of the focus points of the ellipse.

In your investigation, you were able to draw a circle by keeping the distance from the center point (the thumbtack) the same all the way around. This was the way Copernicus envisioned the orbits of the planets in their path around the sun.

In your investigation, you were also able to draw an ellipse by using two stationary tacks. The position of each tack is known as a focus point. All ellipses have two focus points, but neither of them is in the center of the ellipse.

Kepler's observations, precise measurements, and mathematical calculations convinced him that the planets did move around the sun, but he did not agree that their orbits were a perfect circle. He found that the planet Mars orbited the sun in an elliptical orbit with the sun at one of the focus points. (The other focus point doesn't represent anything that can be seen.) The elliptical orbits were shown to be true of all the planets.

The elliptical drawing you made is similar to the way the earth and the planets actually travel around the sun. During a yearlong orbit around the sun, each planet reaches a point when it is closest to the sun and another point when it reaches its greatest distance from the sun.

Don't be confused about the effects of the earth's elliptical orbit. The earth's temperature doesn't become hotter when it is a little closer to the sun. Likewise, the earth's temperature doesn't become colder when it is a little farther away from the sun. Seasons occur because of the way the earth is tilted on its axis and not because of its elliptical orbit. The sun's rays are hotter and more direct during the summer. They are cooler and more slanted during the winter.

Kepler spent years wondering why his observations about Mars did not exactly agree with his calculations for circular orbits around the sun. Some of Kepler's friends tried to convince him that the differences weren't that great, and they weren't important. However, Kepler persisted in trying to figure out how the planets traveled in their orbits around the sun, because he believed that God had created an orderly and predictable universe.

By the end of his life, Kepler had discovered three basic laws about how planets orbit the sun. He is probably most remembered for discovering that planets move around the sun in predictable elliptical orbits. For some reason, fellow astronomers did not consider Kepler's ideas very important at first. However, his three laws became the foundation for astronomy and are still found in modern textbooks.

20

Making Connections

Copernicus was an astronomer who cautiously proposed that the planets orbited the sun. The stars were very predictable and certain constellations could always be seen at the expected seasons. However, the movement of the planets could not be explained in any logical way if they orbited the earth. Their movements made perfect sense if they orbited the sun. This idea caused a great deal of controversy at first, but Kepler's mathematical calculations explained the motion of the planets so precisely, that eventually there could be no doubt that they orbited the sun.

The outer planets travel around the sun in more elongated elliptical orbits than the rocky planets. In fact, the orbits of Neptune and Pluto sometimes cross each other because of their elongated orbits. Comets also orbit the sun, and the shapes of their orbits are even more elongated than any of the planets.

Tycho Brahe
1546–1601

Dig Deeper

- Write about Kepler's early life and his religious beliefs. Tell about the hardships he had to overcome as an adult.
- Write a short story about Tycho Brahe's life. What were some of the things that he contributed to the field of science? What were some ideas that he got wrong?

What Did You Learn?

1. Before the time of Copernicus, scientists believed that the sun, the stars, the moon, and the planets orbited the earth. Explain what Copernicus proposed that disagreed with this.
2. Copernicus believed that the planets orbited the sun in perfect circles. What kind of orbit did Kepler propose that planets followed?
3. What is an elliptical orbit?
4. The earth gets a little closer to the sun at one phase of its orbit. Does the earth get warmer when it is closer to the sun?
5. What causes the different seasons of the earth?
6. What is the difference in the orbits of the inner planets and the outer planets?
7. Do comets orbit the sun? What kind of orbits do they follow?
8. Kepler discovered three basic laws about planetary motion in the 1600s. Do modern astronomers still use these same laws as they study the planets?
9. Did most scientists immediately agree with Copernicus when he proposed that the planets revolve in orbits around the sun?

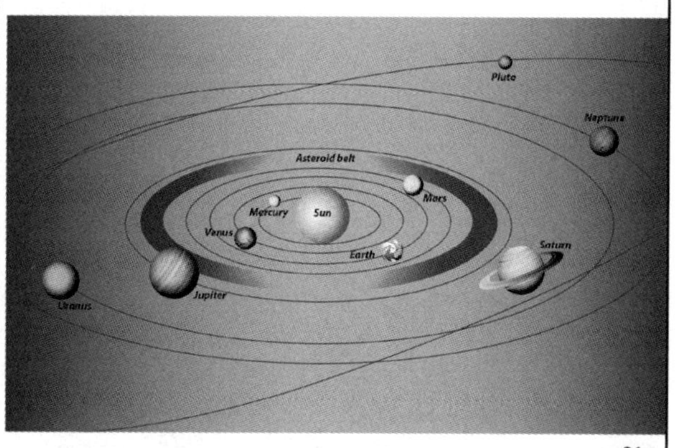

21

5. What causes the different seasons of the earth? *The different seasons of the earth are caused by the tilt of the earth on its axis. The sun's rays are hotter and more direct during the summer. They are cooler and more slanted during the winter.*

6. What is the difference in the orbits of the inner planets and the outer planets? *The outer planets travel around the sun in more elongated elliptical orbits than the rocky planets.*

7. Do comets orbit the sun? What kind of orbits do they follow? *Comets orbit the sun, but their orbits are extremely elliptical.*

8. Kepler discovered three basic laws about planetary motion in the 1600s. Do modern astronomers still use these same laws as they study the planets? *Yes, they are one of the foundations for astronomy today.*

9. Did most scientists immediately agree with Copernicus when he proposed that the planets revolve in orbits around the sun? *No, many scientists had a hard time accepting Copernicus' ideas, because it looked to them like the sun and the stars were moving and the earth was not.*

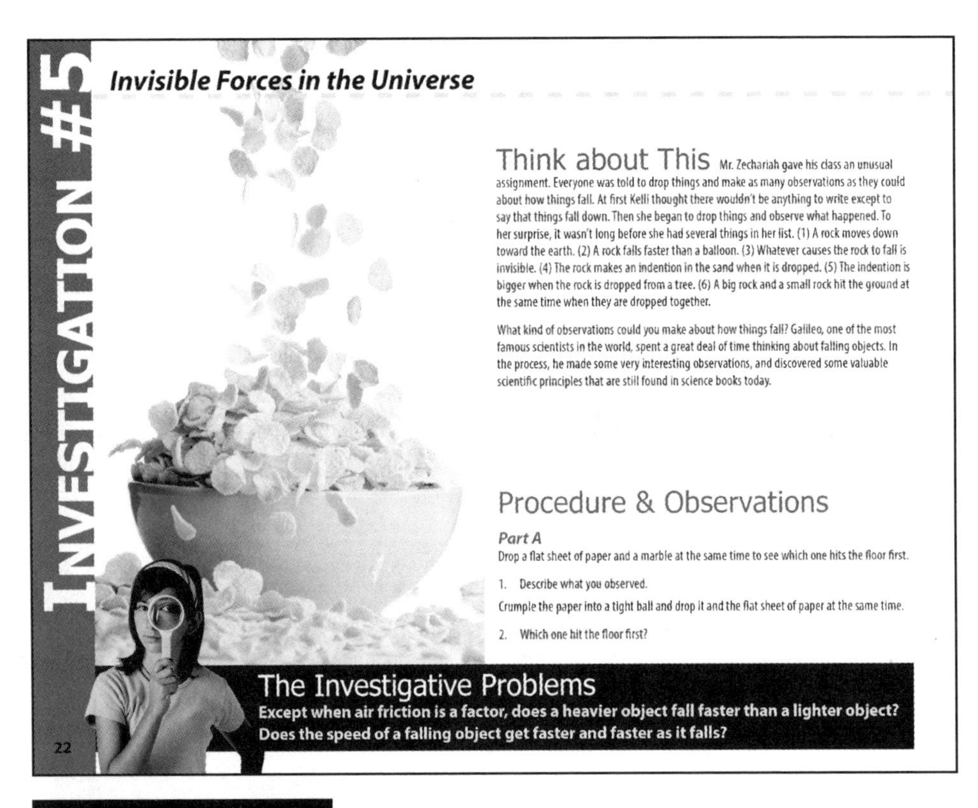

Investigation #5

Invisible Forces in the Universe

Think about This

Mr. Zechariah gave his class an unusual assignment. Everyone was told to drop things and make as many observations as they could about how things fall. At first Kelli thought there wouldn't be anything to write except to say that things fall down. Then she began to drop things and observe what happened. To her surprise, it wasn't long before she had several things in her list. (1) A rock moves down toward the earth. (2) A rock falls faster than a balloon. (3) Whatever causes the rock to fall is invisible. (4) The rock makes an indention in the sand when it is dropped. (5) The indention is bigger when the rock is dropped from a tree. (6) A big rock and a small rock hit the ground at the same time when they are dropped together.

What kind of observations could you make about how things fall? Galileo, one of the most famous scientists in the world, spent a great deal of time thinking about falling objects. In the process, he made some very interesting observations, and discovered some valuable scientific principles that are still found in science books today.

Procedure & Observations

Part A

Drop a flat sheet of paper and a marble at the same time to see which one hits the floor first.

1. Describe what you observed.

Crumple the paper into a tight ball and drop it and the flat sheet of paper at the same time.

2. Which one hit the floor first?

The Investigative Problems

Except when air friction is a factor, does a heavier object fall faster than a lighter object? Does the speed of a falling object get faster and faster as it falls?

22

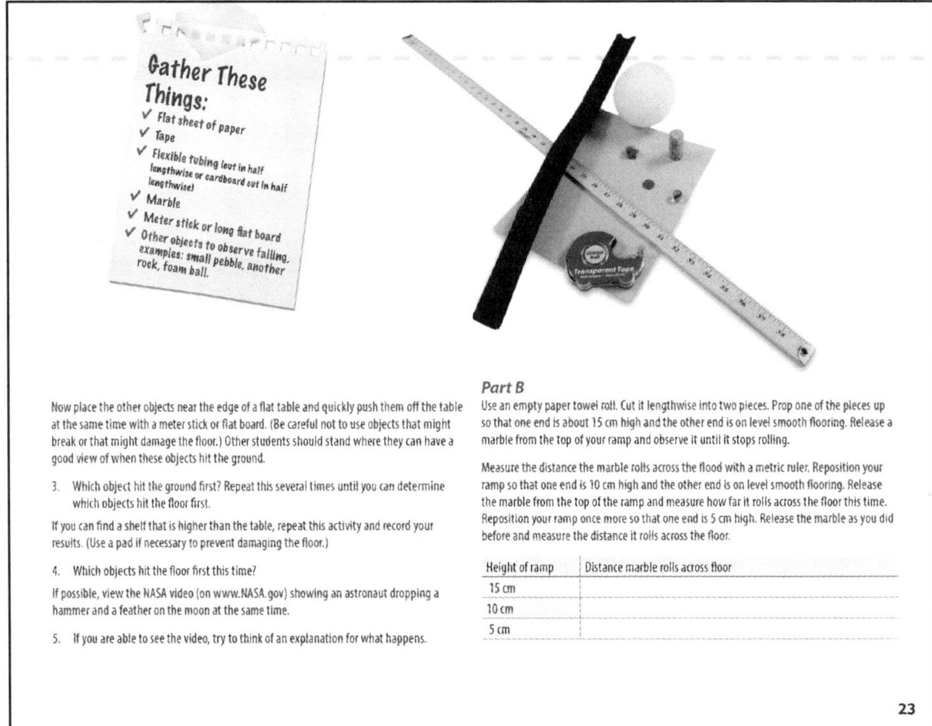

Gather These Things:

- ✓ Flat sheet of paper
- ✓ Tape
- ✓ Flexible tubing (cut in half lengthwise or cardboard cut in half lengthwise)
- ✓ Marble
- ✓ Meter stick or long flat board
- ✓ Other objects to observe falling, examples: small pebble, another rock, foam ball.

Now place the other objects near the edge of a flat table and quickly push them off the table at the same time with a meter stick or flat board. (Be careful not to use objects that might break or that might damage the floor.) Other students should stand where they can have a good view of when these objects hit the ground.

3. Which object hit the ground first? Repeat this several times until you can determine which objects hit the floor first.

If you can find a shelf that is higher than the table, repeat this activity and record your results. (Use a pad if necessary to prevent damaging the floor.)

4. Which objects hit the floor first this time?

If possible, view the NASA video (on www.NASA.gov) showing an astronaut dropping a hammer and a feather on the moon at the same time.

5. If you are able to see the video, try to think of an explanation for what happens.

Part B

Use an empty paper towel roll. Cut it lengthwise into two pieces. Prop one of the pieces up so that one end is about 15 cm high and the other end is on level smooth flooring. Release a marble from the top of your ramp and observe it until it stops rolling.

Measure the distance the marble rolls across the flood with a metric ruler. Reposition your ramp so that one end is 10 cm high and the other end is on level smooth flooring. Release the marble from the top of the ramp and measure how far it rolls across the floor this time. Reposition your ramp once more so that one end is 5 cm high. Release the marble as you did before and measure the distance it rolls across the floor.

Height of ramp	Distance marble rolls across floor
15 cm	
10 cm	
5 cm	

23

Objectives

1. A falling object is pulled down by a gravitational force. Air friction exerts an upward force on a falling object.
2. The speed of a falling object does not depend on its size and weight unless air friction is a factor.
3. Falling objects on earth (and other places) accelerate as they fall. They accelerate at a predictable rate.
4. Galileo studied acceleration by rolling balls down ramps and measuring the distance they moved each second of time.
5. Galileo was one of the first scientists to insist that scientific ideas and explanations need to be tested.

Note

The next investigation calls for a flexible tube that is cut lengthwise into two pieces. Students may use the same tubing for this investigation if they prefer.

What Did You Learn?

1. Why did a flat sheet of paper fall more slowly that an equal size wadded-up sheet of paper? *Even though gravity pulled down on the paper, air resistance (friction) pushed up on it as it fell through the air. The flat sheet of paper was affected more by air resistance than the wadded-up sheet of paper.*
2. When air friction isn't a major factor on a falling object, do all objects dropped at the same time, and from the same height, hit the ground at the same time? *Yes*
3. An astronaut dropped a feather and a hammer on the moon at the same time from the same height. Why did they hit the ground at the same time? *The moon doesn't have an atmosphere of air around it, so there was no air resistance to push up on the feather.*

The Science Stuff

When an object falls through the air, gravity exerts a force on it that pulls the object down. Air resistance (friction) pushes up on objects that fall through the air. The flat sheet of paper, and possibly some of your other objects, did not fall as fast as other objects. These things were affected by the force of air friction.

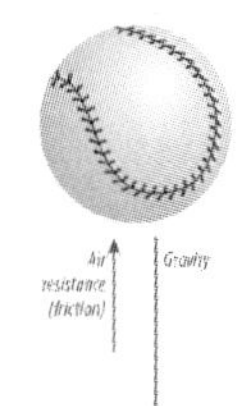

Except for the items where air friction is a noticeable force on them, you should have observed that the other objects all hit the floor at the same time. If air friction is not a major influence on falling objects, all objects will fall at the same rate. If these objects are dropped from the same level at the same time, they will hit the ground at the same time. You might think that a large heavy rock would fall faster than a smaller pebble, but that is not the case. The rate at which objects fall does not depend on their size or weight (except when affected by air friction).

Galileo spent a lot of time investigating and thinking about gravity and how things fall. He knew that gravity is a key force that acts on falling objects. He recognized that friction gets in the way and other objects can get in the way.

A philosopher named Aristotle, who lived before the time of Christ, had also studied many of the same things Galileo was investigating. Galileo thought that some of Aristotle's ideas were wrong, even though they had been accepted as true for hundreds of years. Aristotle made many conclusions based on what seemed logical to him, but Galileo knew that having a logical idea was not enough. Galileo was one of the first scientists to insist that ideas and explanations need to be tested.

Galileo wanted to take some measurements of falling objects, but they moved too fast to tell what was happening. He decided to roll balls down ramps to slow their motion enough to study how far they moved during different time intervals. The lower ramps allowed the balls to roll slow enough to take some measurements.

As long as the force of gravity kept being applied to the ball, the ball continued to accelerate or roll faster and faster.

Galileo made careful measurements and found a way to keep up with the time as objects rolled down a ramp. He was amazed at what he found. A ball would roll on a smooth ramp 1 unit during the 1st second; 3 during the 2nd second; 5 during the 3rd second; and 7 during the 4th second. When he added the total distances the ball rolled during each second, he found this pattern:

Table 5.1

Time	Present & Past Distances	Total Distance Covered
1 sec	1	1 (1 x 1)
2 sec	1 + 3	4 (2 x 2)
3 sec	1 + 3 + 5	9 (3 x 3)
4 sec	1 + 3 + 5 + 7	16 (4 x 4)

At the end of the first second, a ball would have rolled 1 unit. At the end of the next second, the ball would have rolled a total of 4 units. At the end of the third second, the ball would have rolled a total of 9 units. These patterns continued as long as Galileo could measure them.

If you are good at math, you may have observed a really interesting mathematical pattern, just as Galileo did. The total distance covered from the time the ball began rolling is a squared number. These patterns are so obvious, that you should be able to complete the distances for the next several seconds.

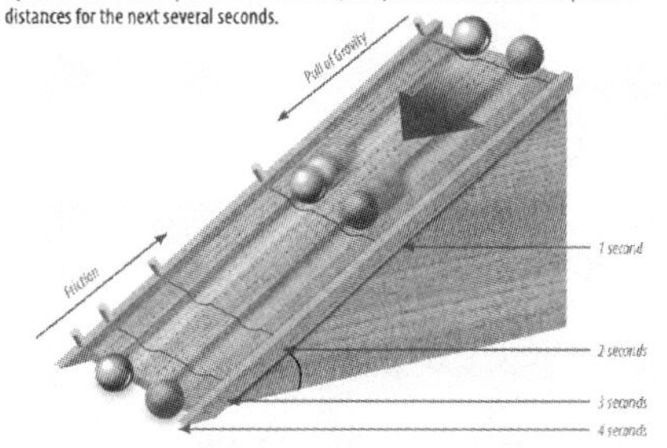

Making Connections

Isaac Newton was born the same year that Galileo died, but Newton was fascinated by the kind of research Galileo did about gravity. Newton continued to expand on Galileo's research. He realized that gravity exists throughout the universe and follows the same "rules" everywhere. The amount of gravitational attraction that exists between two objects depends on two things: their masses (weight) and the distance between them. Newton found a mathematical formula that could be used to calculate the gravitational attraction anywhere in the universe. This became known as the law of universal gravitation.

Dig Deeper

We know many things about gravity, forces, and motion. We know very little about how or why gravity makes things fall. Try to find some explanations for why gravity causes things to fall as they do. Many of the explanations are based on complicated math, but try to summarize at least one of the main possible explanations.

Make a time-line, or add to a time-line you already have, the names of Galileo, Tycho Brahe, Johannes Kepler, Nicholas Copernicus, and Isaac Newton. Include their birth and death years. Galileo was one of the first scientists to use a telescope to study the planets. Write "early telescopes" around the years 1608–1610. Include other historical events with their dates. Now add the name of Albert Einstein, with his birth and death years, at the proper place in the time-line. The things discovered by these earlier scientists were expanded by Einstein's general theory of relativity and his special theory of relativity to help explain the forces of our universe.

What Did You Learn?

1. Why did a flat sheet of paper fall more slowly that an equal size wadded-up sheet of paper?
2. When air friction isn't a major factor on a falling object, do all objects dropped at the same time, and from the same height, hit the ground at the same time?
3. An astronaut dropped a feather and a hammer on the moon at the same time from the same height. Why did they hit the ground at the same time?
4. Around what time period did scientists begin to use telescopes to study planets, moons, and other objects in space?
5. The amount of gravitational attraction that exists between objects depends on what two things?
6. What term is used to describe when an object is moving and it keeps on getting faster and faster?
7. Is this true or false? Galileo noted that as long as the force of gravity kept being applied to a rolling ball on a ramp, the ball continued to accelerate.
8. Is this true or false? Galileo knew that friction and other things can get in the way of how falling objects move and cause them to slow down or stop.
9. Aristotle and Galileo were both scientists, although Aristotle lived hundreds of years before Galileo. Why was Galileo a better scientist that Aristotle?
10. Look at the numbers in Table 5.1 showing the distance a ball rolled each second. See if you can complete the line for five seconds on the Table, based on the patterns for the first four seconds.

4. Around what time period did scientists begin to use telescopes to study planets, moons, and other objects in space? *During the early 1600s*

5. The amount of gravitational attraction that exists between objects depends on what two things? *Their masses and the distance between them*

6. What term is used to describe when an object is moving and it keeps on getting faster and faster? *Acceleration (or positive acceleration)*

7. Is this true or false? Galileo noted that as long as the force of gravity kept being applied to a rolling ball on a ramp, the ball continued to accelerate. *True*

8. Is this true or false? Galileo knew that friction and other things can get in the way of how falling objects move and cause them to slow them down or stop. *True*

9. Aristotle and Galileo were both scientists, although Aristotle lived hundreds of years before Galileo. Why was Galileo a better scientist that Aristotle? *Aristotle made many conclusions based on what seemed logical to him, but Galileo knew that having a logical idea was not enough. Galileo insisted that ideas and explanations need to be tested.*

10. Look at the numbers in Table 5.1 showing the distance a ball rolled each second. See if you can complete the line for five seconds on the Table, based on the patterns for the first four seconds. *5 sec* *1 + 3 + 5 + 7 + 9* *25* *(5 x 5)*

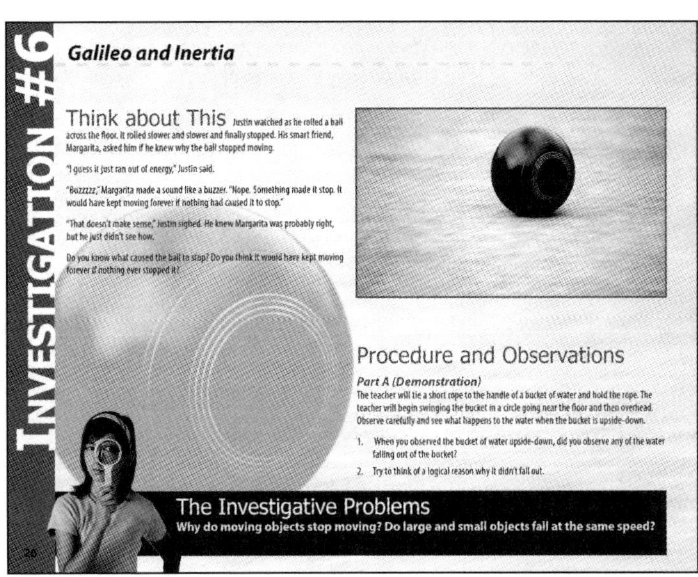

INVESTIGATION #6

Galileo and Inertia

Think about This Justin watched as he rolled a ball across the floor. It rolled slower and slower and finally stopped. His smart friend, Margarita, asked him if he knew why the ball stopped moving.

"I guess it just ran out of energy," Justin said.

"Buzzzzz," Margarita made a sound like a buzzer. "Nope. Something made it stop. It would have kept moving forever if nothing had caused it to stop."

"That doesn't make sense," Justin sighed. He knew Margarita was probably right, but he just didn't see how.

Do you know what caused the ball to stop? Do you think it would have kept moving forever if nothing ever stopped it?

Procedure and Observations

Part A (Demonstration)

The teacher will tie a short rope to the handle of a bucket of water and hold the rope. The teacher will begin swinging the bucket in a circle going near the floor and then overhead. Observe carefully and see what happens to the water when the bucket is upside-down.

1. When you observed the bucket of water upside-down, did you observe any of the water falling out of the bucket?
2. Try to think of a logical reason why it didn't fall out.

The Investigative Problems

Why do moving objects stop moving? Do large and small objects fall at the same speed?

26

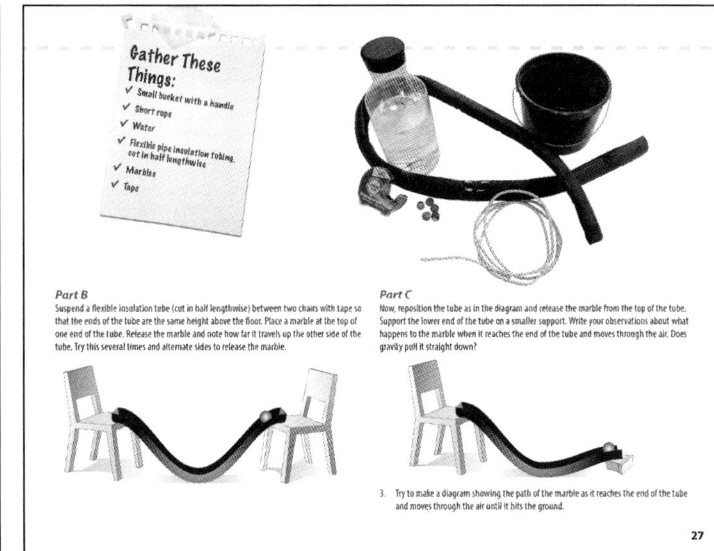

Gather These Things:

- ✓ Small bucket with a handle
- ✓ Short rope
- ✓ Water
- ✓ Flexible pipe insulation tubing, cut in half lengthwise
- ✓ Marbles
- ✓ Tape

Part B

Suspend a flexible insulation tube (cut in half lengthwise) between two chairs with tape so that the ends of the tube are the same height above the floor. Place a marble at the top of one end of the tube. Release the marble and note how far it travels up the other side of the tube. Try this several times and alternate sides to release the marble.

Part C

Now, reposition the tube as in the diagram and release the marble from the top of the tube. Support the lower end of the tube on a smaller support. Write your observations about what happens to the marble when it reaches the end of the tube and moves through the air. Does gravity pull it straight down?

3. Try to make a diagram showing the path of the marble as it reaches the end of the tube and moves through the air until it hits the ground.

27

Objectives

1. Galileo recognized that once an object starts to move, it will continue to move without being pushed along, unless another force interferes with it.
2. He recognized that if an object is not moving it has a tendency to resist moving.
3. This is a property of all matter and is known as inertia.
4. The gravitational pull on a falling object is balanced by the object's resistance to moving, causing a small object to fall at the same rate as a large object
5. Newton formulated three laws of motion and the universal law of gravitation by continuing to study Galileo's ideas.

Note

Another interesting investigation is to place two marbles in the middle of the tubing. Let one marble roll down the tubing and observe what happens. Let one marble roll down the opposite side of the tubing and observe what happens. The rolling marble will come to a stop, but the other marbles will begin to move.

1. In the activity you did, a marble rolled down a piece of tubing. What force caused it to roll down? *Gravity*
2. In this same activity, the marble continued to roll and moved up the tubing for a distance. What property of matter cause it to keep rolling up a hill? *Inertia*
3. What effect did the force of friction have on the marble's motion? *It pushed the opposite way the marble was moving.*

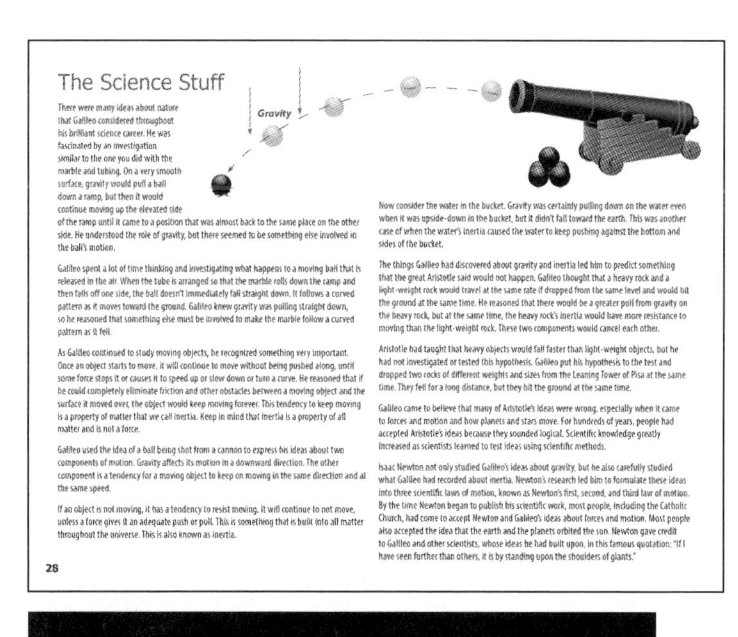

The Science Stuff

There were many ideas about nature that Galileo considered throughout his brilliant science career. He was fascinated by an investigation similar to the one you did with the marble and tubing. On a very smooth surface, gravity would pull a ball down a ramp, but then it would continue moving up the elevated side of the ramp until it came to a position that was almost back to the same place on the other side. He understood the role of gravity, but there seemed to be something else involved in the ball's motion.

Galileo spent a lot of time thinking and investigating what happens to a moving ball that is released in the air. When the tube is arranged so that the marble rolls down the ramp and then falls off one side, the ball doesn't immediately fall straight down. It follows a curved pattern as it moves toward the ground. Galileo knew gravity was pulling straight down, so he reasoned that something else must be involved to make the marble follow a curved pattern as it fell.

As Galileo continued to study moving objects, he recognized something very important. Once an object starts to move, it will continue to move without being pushed along, until some force stops it or causes it to speed up or slow down or turn a curve. He reasoned that if he could completely eliminate friction and other obstacles between a moving object and the surface it moved over, the object would keep moving forever. This tendency to keep moving is a property of matter that we call inertia. Keep in mind that inertia is a property of all matter and is not a force.

Galileo used the idea of a ball being shot from a cannon to express his ideas about two components of motion. Gravity affects its motion in a downward direction. The other component is a tendency for a moving object to keep on moving in the same direction and at the same speed.

If an object is not moving, it has a tendency to resist moving. It will continue to not move, unless a force gives it an adequate push or pull. This is something that is built into all matter throughout the universe. This is also known as inertia.

Now consider the water in the bucket. Gravity was certainly pulling down on the water even when it was upside-down in the bucket, but it didn't fall toward the earth. This was another case of when the water's inertia caused the water to keep pushing against the bottom and sides of the bucket.

The things Galileo had discovered about gravity and inertia led him to predict something that the great Aristotle said would not happen. Galileo thought that a heavy rock and a light-weight rock would travel at the same rate if dropped from the same level and would hit the ground at the same time. He reasoned that there would be a greater pull from gravity on the heavy rock, but at the same time, the heavy rock's inertia would have more resistance to moving than the light-weight rock. These two components would cancel each other.

Aristotle had taught that heavy objects would fall faster than light-weight objects, but he had not investigated or tested this hypothesis. Galileo put his hypothesis to the test and dropped two rocks of different weights and sizes from the Leaning Tower of Pisa at the same time. They fell for a long distance, but they hit the ground at the same time.

Galileo came to believe that many of Aristotle's ideas were wrong, especially when it came to forces and motion and how planets and stars move. For hundreds of years, people had accepted Aristotle's ideas because they sounded logical. Scientific knowledge greatly increased as scientists learned to test ideas using scientific methods.

Isaac Newton not only studied Galileo's ideas about gravity, but he also carefully studied what Galileo had recorded about inertia. Newton's research led him to formulate these ideas into three scientific laws of motion, known as Newton's first, second, and third law of motion. By the time Newton began to publish his scientific work, most people, including the Catholic Church, had come to accept Newton and Galileo's ideas about forces and motion. Most people also accepted the idea that the earth and the planets orbited the sun. Newton gave credit to Galileo and other scientists, whose ideas he had built upon, in this famous quotation: "If I have seen further than others, it is by standing upon the shoulders of giants."

28

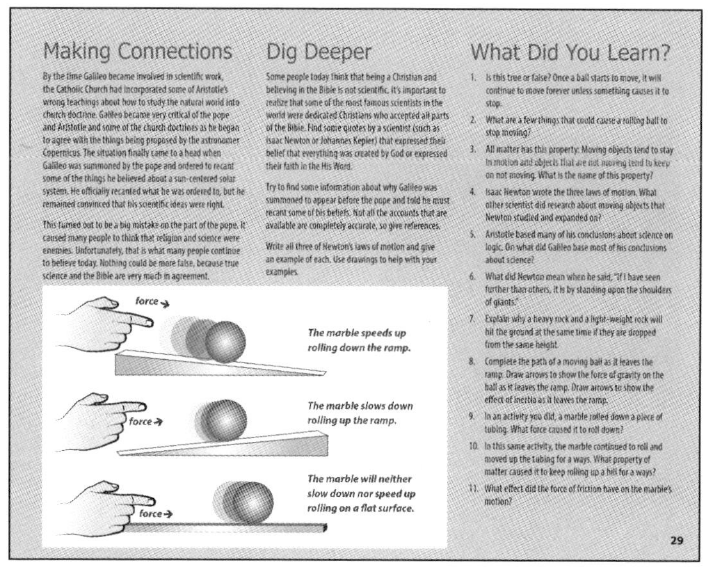

Making Connections

By the time Galileo became involved in scientific work, the Catholic Church had incorporated some of Aristotle's wrong teachings about how to study the natural world into church doctrine. Galileo became very critical of the pope and Aristotle and some of the church doctrines as he began to agree with the things being proposed by the astronomer Copernicus. The situation finally came to a head when Galileo was summoned by the pope and ordered to recant some of the things he believed about a sun-centered solar system. He officially recanted what he was ordered to, but he remained convinced that his scientific ideas were right.

This turned out to be a big mistake on the part of the pope. It caused many people to think that religion and science were enemies. Unfortunately, that is what many people continue to believe today. Nothing could be more false, because true science and the Bible are very much in agreement.

Dig Deeper

Some people today think that being a Christian and believing in the Bible is not scientific. It's important to realize that some of the most famous scientists in the world were dedicated Christians who accepted all parts of the Bible. Find some quotes by a scientist (such as Isaac Newton or Johannes Kepler) that expressed their belief that everything was created by God or expressed their faith in the His Word.

Try to find some information about why Galileo was summoned to appear before the pope and told he must recant some of his beliefs. Not all the accounts that are available are completely accurate, so give references.

Write all three of Newton's laws of motion and give an example of each. Use drawings to help with your examples.

What Did You Learn?

1. Is this true or false? Once a ball starts to move, it will continue to move forever unless something causes it to stop.
2. What are a few things that could cause a rolling ball to stop moving?
3. All matter has this property: Moving objects tend to stay in motion and objects that are not moving tend to keep on not moving. What is the name of this property?
4. Isaac Newton wrote the three laws of motion. What other scientist did research about moving objects that Newton studied and expanded on?
5. Aristotle based many of his conclusions about science on logic. On what did Galileo base most of his conclusions about science?
6. What did Newton mean when he said, "If I have seen further than others, it is by standing upon the shoulders of giants."
7. Explain why a heavy rock and a light-weight rock will hit the ground at the same time if they are dropped from the same height.
8. Complete the path of a moving ball as it leaves the ramp. Draw arrows to show the force of gravity on the ball as it leaves the ramp. Draw arrows to show the effect of inertia as it leaves the ramp.
9. In an activity you did, a marble rolled down a piece of tubing. What force caused it to roll down?
10. In this same activity, the marble continued to roll and moved up the tubing for a ways. What property of matter caused it to keep rolling up a hill for a ways?
11. What effect did the force of friction have on the marble's motion?

29

What Did You Learn?

1. Is this true or false? Once a ball starts to move, it will continue to move forever unless something causes it to stop. *Yes*
2. What are a few things that could cause a rolling ball to stop moving? *Some kind of force could stop it, such as friction or a collision with another object.*
3. All matter has this property: Moving objects tend to stay in motion and objects that are not moving tend to keep on not moving. What is the name of this property? *Inertia*
4. Isaac Newton wrote the three laws of motion. What other scientist did research about moving objects that Newton studied and expanded on? *Galileo*
5. Aristotle based many of his conclusions about science on logic. On what did Galileo base most of his conclusions about science? *Scientific tests*
6. What did Newton mean when he said, "If I have seen further than others, it is by standing upon the shoulders of giants." *He studied the research that other scientists had already done and then added to what they had learned.*
7. Explain why a heavy rock and a light-weight rock will hit the ground at the same time if they are dropped from the same height. *There is a greater pull from gravity on the heavy rock, but at the same time, the heavy rock's inertia has more resistance to moving than the light-weight rock. These two components cancel each other.*
8. Complete the path of a moving ball as it leaves the ramp. Draw arrows to show the force of gravity on the ball as it leaves the ramp. Draw arrows to show the effect of inertia as it leaves the ramp.
9. In the activity you did, a marble rolled down a piece of tubing. What force caused it to roll down? *Gravity*
10. In this same activity, the marble continued to roll and moved up the tubing for a distance. What property of matter caused it to keep rolling up a hill? *Inertia*
11. What effect did the force of friction have on the marble's motion? *It pushed the opposite way the marble was moving.*

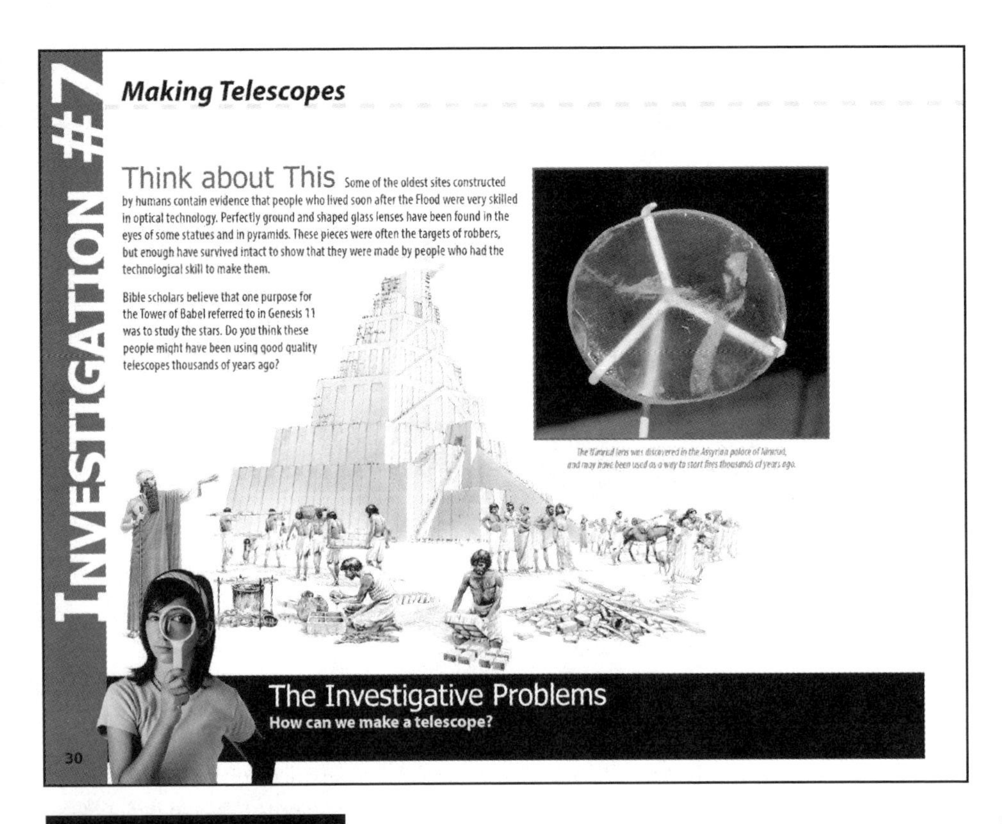
INVESTIGATION #7

Making Telescopes

Think about This Some of the oldest sites constructed by humans contain evidence that people who lived soon after the Flood were very skilled in optical technology. Perfectly ground and shaped glass lenses have been found in the eyes of some statues and in pyramids. These pieces were often the targets of robbers, but enough have survived intact to show that they were made by people who had the technological skill to make them.

Bible scholars believe that one purpose for the Tower of Babel referred to in Genesis 11 was to study the stars. Do you think these people might have been using good quality telescopes thousands of years ago?

The Nimrud lens was discovered in the Assyrian palace of Nimrud, and may have been used as a way to start fires thousands of years ago.

The Investigative Problems

How can we make a telescope?

30

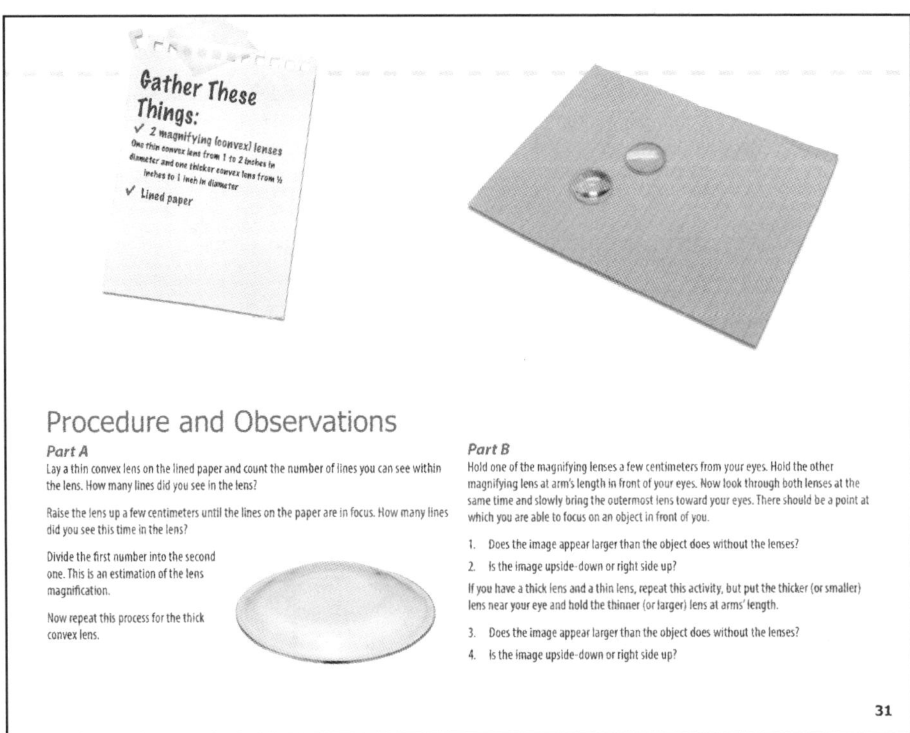
Gather These Things:

✓ 2 magnifying (convex) lenses
One thin convex lens from 1 to 2 inches in diameter and one thicker convex lens from ½ inches to 1 inch in diameter

✓ Lined paper

Procedure and Observations

Part A

Lay a thin convex lens on the lined paper and count the number of lines you can see within the lens. How many lines did you see in the lens?

Raise the lens up a few centimeters until the lines on the paper are in focus. How many lines did you see this time in the lens?

Divide the first number into the second one. This is an estimation of the lens magnification.

Now repeat this process for the thick convex lens.

Part B

Hold one of the magnifying lenses a few centimeters from your eyes. Hold the other magnifying lens at arm's length in front of your eyes. Now look through both lenses at the same time and slowly bring the outermost lens toward your eyes. There should be a point at which you are able to focus on an object in front of you.

1. Does the image appear larger than the object does without the lenses?
2. Is the image upside-down or right side up?

If you have a thick lens and a thin lens, repeat this activity, but put the thicker (or smaller) lens near your eye and hold the thinner (or larger) lens at arms' length.

3. Does the image appear larger than the object does without the lenses?
4. Is the image upside-down or right side up?

31

OBJECTIVES

1. Convex lenses are thicker in the middle and thinner on the edges. Convex lenses bend light as it passes through. These lenses are known as magnifying lenses.

2. Telescopes are made of one thick convex lens in the eyepiece and one thinner, wider lens for the second lens. The combination of the two lenses produces an image that is larger and upside-down.

NOTE

It is very likely that people who lived during pre-Flood times had discovered how to make high-quality lenses and how to use them in telescopes. They had probably developed other kinds of technology and had made scientific discoveries as well. The people who survived the Flood could have passed along the knowledge they had to their children. Over the years, some of this technology would have been lost, but was rediscovered during the 1600s.

Archeologists have discovered some very high-tech artifacts from civilizations thought to be thousands of years old. These artifacts have caused some people to believe that intelligent space aliens are responsible for showing men how to build these technologies. Another more logical conclusion is that these artifacts survived the Flood or were created soon after the Flood by an intelligent group of people.

For more information, you can find these resources: *Frozen in Time; Uncovering the Mysterious Woolly Mammoth;* and *Genius of Ancient Man.*

If students want to make their own telescopes, paper towel rolls or gift wrapping rolls make good tubes. They can cut one tube lengthwise, overlap the cut edges a little, and then tape the edges securely. This should make the diameter small enough to fit inside the uncut tube and slide back and forth.

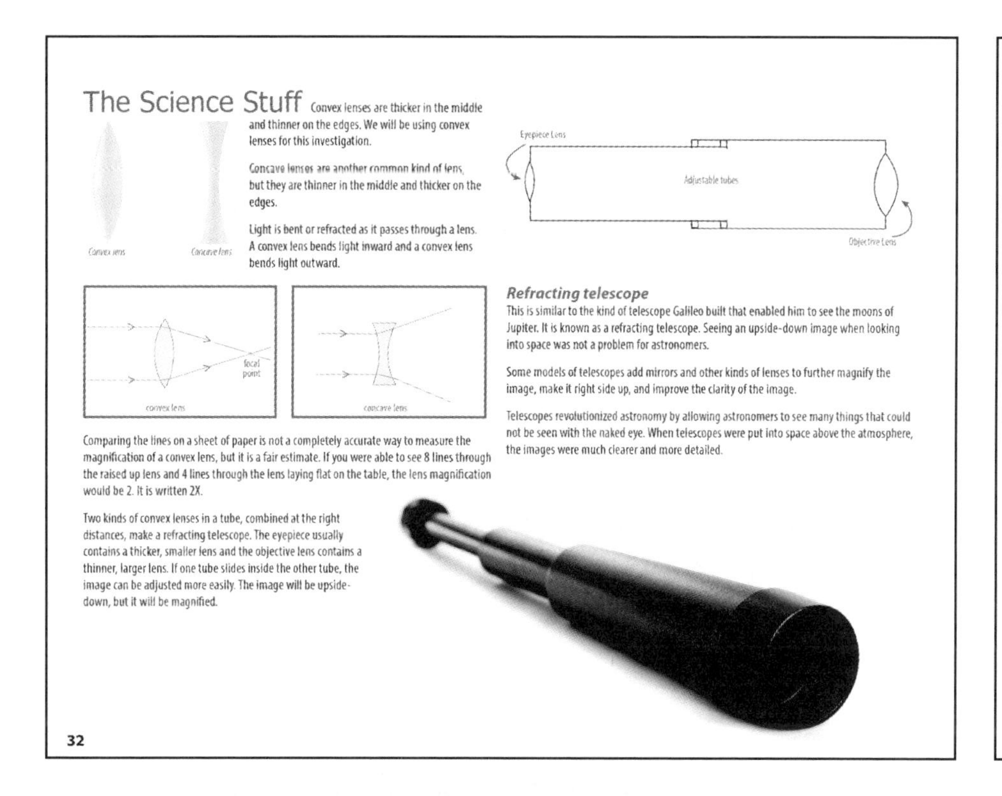

The Science Stuff

Convex lenses are thicker in the middle and thinner on the edges. We will be using convex lenses for this investigation.

Concave lenses are another common kind of lens, but they are thinner in the middle and thicker on the edges.

Light is bent or refracted as it passes through a lens. A convex lens bends light inward and a convex lens bends light outward.

Comparing the lines on a sheet of paper is not a completely accurate way to measure the magnification of a convex lens, but it is a fair estimate. If you were able to see 8 lines through the raised up lens and 4 lines through the lens laying flat on the table, the lens magnification would be 2. It is written 2X.

Two kinds of convex lenses in a tube, combined at the right distances, make a refracting telescope. The eyepiece usually contains a thicker, smaller lens and the objective lens contains a thinner, larger lens. If one tube slides inside the other tube, the image can be adjusted more easily. The image will be upside-down, but it will be magnified.

Refracting telescope

This is similar to the kind of telescope Galileo built that enabled him to see the moons of Jupiter. It is known as a refracting telescope. Seeing an upside-down image when looking into space was not a problem for astronomers.

Some models of telescopes add mirrors and other kinds of lenses to further magnify the image, make it right side up, and improve the clarity of the image.

Telescopes revolutionized astronomy by allowing astronomers to see many things that could not be seen with the naked eye. When telescopes were put into space above the atmosphere, the images were much clearer and more detailed.

32

Making Connections

When the thick lens and the thin lens is reversed and fitted into a tube, a microscope can be made.

Curved mirrors can make objects look smaller or larger. Reflecting telescopes use curved mirrors that create a magnified image and then use a convex lens to magnify the image even more.

Dig Deeper

Make a working refracting telescope by mounting the proper lenses inside two cardboard tubes. Be sure the tubes are long enough and that one tube slides snugly inside the other. Do a search of the Internet or find a good reference book for more detailed instructions.

Try to find more information about ancient optical lenses.

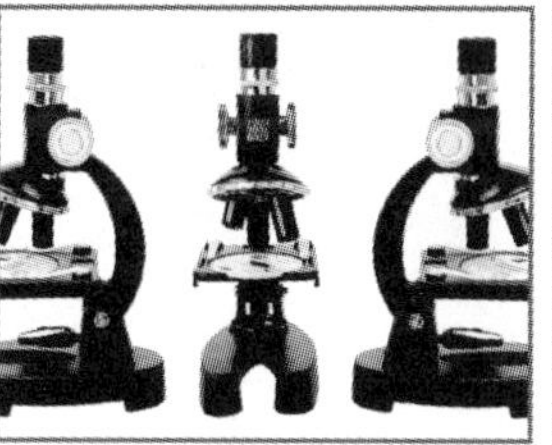

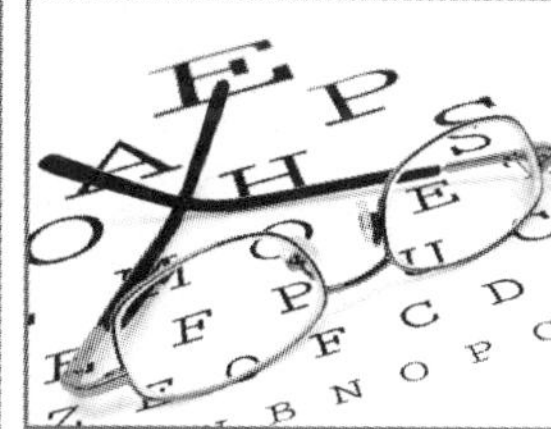

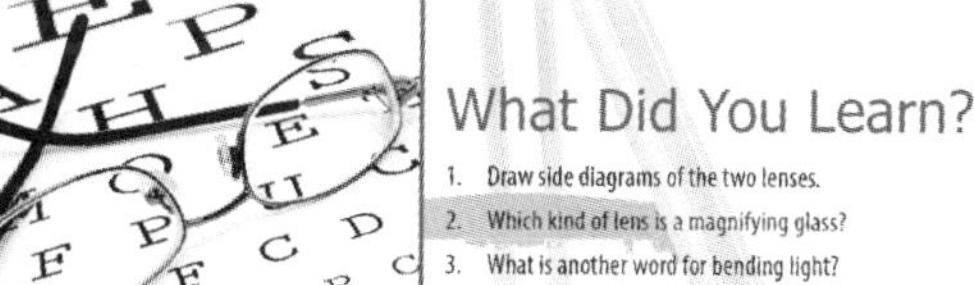

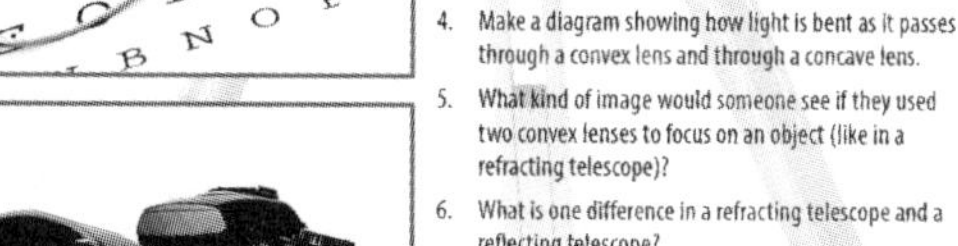

What Did You Learn?

1. Draw side diagrams of the two lenses.
2. Which kind of lens is a magnifying glass?
3. What is another word for bending light?
4. Make a diagram showing how light is bent as it passes through a convex lens and through a concave lens.
5. What kind of image would someone see if they used two convex lenses to focus on an object (like in a refracting telescope)?
6. What is one difference in a refracting telescope and a reflecting telescope?
7. What famous scientist built a telescope that enabled him to see some of the moons around Jupiter?
8. What is the difference in how a refracting telescope and a microscope are made?

33

What Did You Learn?

1. Draw diagrams of the two lenses (page 32).
2. Which kind of lens is a magnifying glass? *A convex lens*
3. What is another word for bending light? *Refraction*
4. Do creation scientists and evolutionary scientists examine the same facts and observations but come to different conclusions when trying to explain what happened in the past?
5. What kind of image would someone see if they used two convex lenses to focus on an object (like in a refracting telescope)? *The image would be upside-down.*
6. What is one difference in a refracting telescope and a reflecting telescope? *A refracting telescope uses two kinds of convex lenses in a tube, combined at the right distances. The image will be upside-down, but it will be magnified. It uses two convex lenses to bend the light. Reflecting telescopes use curved mirrors that create a magnified image and then use a convex lens to magnify the image even more.*
7. What famous scientist built a telescope that enabled him to see some of the moons around Jupiter? *Galileo*
8. What is the difference in how a refracting telescope and a microscope are made? *They both contain two convex lenses in an adjustable tube. One of the lenses is a thick lens and the other one is a thin lens The lenses are reversed in a microscope.*

INVESTIGATION #8

History of Flight

Think about This

Ceborne Perkins and Ora Mae Thomas were born in the United States in 1896. They were fascinated with hot-air balloons, gliders, and air ships as they grew up. They had heard about a glider with an engine on it when they were about 7, but they were 14 before they actually saw a flying machine in the air.

After the United States entered World War I in 1917, newspapers reported that airplanes were being used in the war to drop bombs. An airplane flying over their community early one morning caused quite a stir, because they weren't sure if it was a friendly American plane or an enemy plane that might drop a bomb. Ceborne joined the army that same year and had his first plane ride after his military unit arrived in Europe. He and Ora Mae were married after Ceborne came back to the United States.

By the time the United States entered World War II in 1941, they had children of their own. Their oldest son joined the Air Force and was a pilot until the war ended in 1945. U.S. air planes were a major factor in this war.

The family grew as grandchildren were born. Then on July 20, 1969, at the age of 73, Ceborne and Ora Mae sat in their living room with their 7-year-old grandson and watched the first astronauts walk on the moon on live TV. They found it hard to believe that the world's first motor-powered airplane was being flown by the Wright brothers when they were 7 years old.

Why do you think people were unable to build flying machines for thousands of years, and in a little less than 70 years, airplanes were crossing the oceans and rocket ships were traveling to the moon?

The Investigative Problems

How were gliders an important step that helped to bring about long-distance air travel?

34

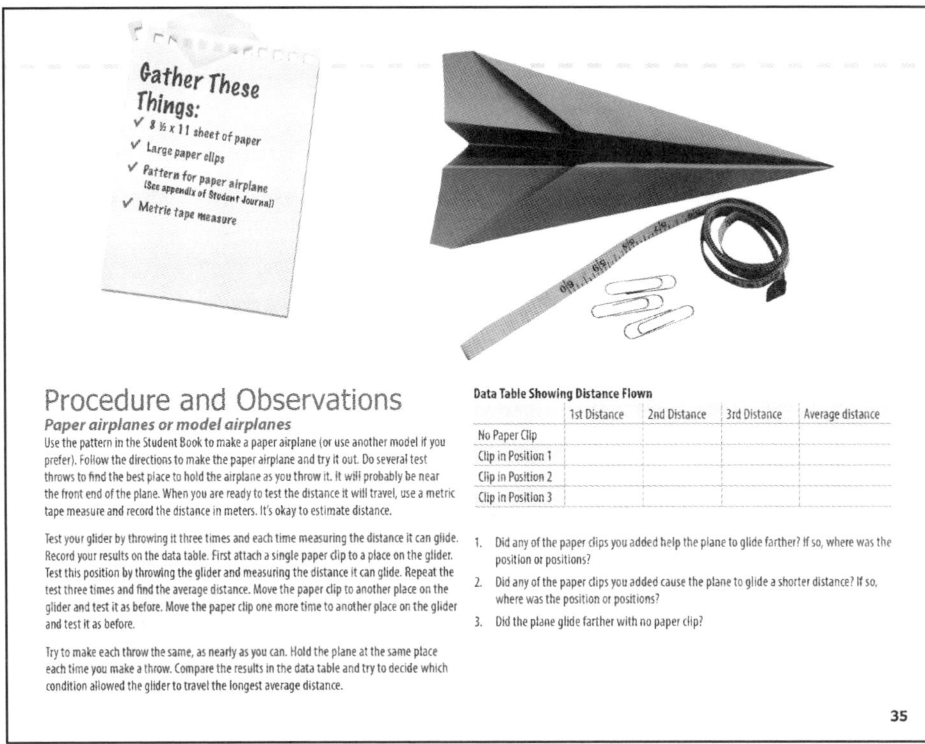

Procedure and Observations

Paper airplanes or model airplanes

Use the pattern in the Student Book to make a paper airplane (or use another model if you prefer). Follow the directions to make the paper airplane and try it out. Do several test throws to find the best place to hold the airplane as you throw it. It will probably be near the front end of the plane. When you are ready to test the distance it will travel, use a metric tape measure and record the distance in meters. It's okay to estimate distance.

Test your glider by throwing it three times and each time measuring the distance it can glide. Record your results on the data table. First attach a single paper clip to a place on the glider. Test this position by throwing the glider and measuring the distance it can glide. Repeat the test three times and find the average distance. Move the paper clip to another place on the glider and test it as before. Move the paper clip one more time to another place on the glider and test it as before.

Try to make each throw the same, as nearly as you can. Hold the plane at the same place each time you make a throw. Compare the results in the data table and try to decide which condition allowed the glider to travel the longest average distance.

Data Table Showing Distance Flown

	1st Distance	2nd Distance	3rd Distance	Average distance
No Paper Clip				
Clip in Position 1				
Clip in Position 2				
Clip in Position 3				

1. Did any of the paper clips you added help the plane to glide farther? If so, where was the position or positions?
2. Did any of the paper clips you added cause the plane to glide a shorter distance? If so, where was the position or positions?
3. Did the plane glide farther with no paper clip?

35

OBJECTIVES

1. Men tried unsuccessfully for centuries to make devices that would fly with a man attached.
2. Hot-air balloons were successfully used beginning in the late 1700s, but they were more of a sport for a few people than a commercial effort.
3. During the early 1900s, a few hydrogen-filled airships were built that could carry a large number of people. Before they could become a popular means of travel, one of the air ships exploded and crashed.
4. During the late 1800s, air gliders were developed and used primarily as a sport.
5. In 1903, the Wright brothers attached a light-weight gasoline engine to a glider and conducted the first motorized flight in history. They made many modifications to their flying machine to make it capable of longer flights.
6. By the time World War II began, airplanes were capable of carrying out war missions.
7. Small rockets have been around for hundreds of years, but during the 1930s and 1940s, large powerful rockets were developed that could travel long distances and carry heavy loads.

NOTE

Another reason for using time-lines is to help students see that scientific discoveries usually come before technological inventions and that improvements occur rapidly after basic designs are achieved. The Wright brothers based their designs on research done by others. Later researchers improved the Wright brothers' airplanes. Rockets depended on understanding Newton's second law of motion, gravity, thrust, and other principles about forces and motion. Simple rockets became three-stage rockets. Combustion in the air was replaced with rockets that carried their own supply of oxygen into space. Current electricity was discovered by Michael Faraday and Joseph Henry. After electric generators were built, a number of electric devices were built. Eventually, computers were designed and built. Computers were essential to space travel.

Bernoulli's principle and the forces that act on airplanes are explained more fully in the book *Forces and Motion,* published by Master Books and authored by Tom DeRosa and Carolyn Reeves. You might find it helpful to review some of these principles with the students.

Consider buying a rocket kit and launching a rocket. Use a timer and see how much times passes between when the rocket is launched and when it hits the ground. *If you do this be sure everyone around uses safety glasses and all safety procedures are followed. (No shortcuts here!)*

The Science Stuff

Men tried for centuries to fly by imitating birds and devising ways to build wings they could flap with their muscles. There were many attempts, often with disastrous results, but this plan never worked. People were too heavy and their muscles were too weak to fly.

The next approach was to rise above the earth using hot-air balloons. This was almost as dangerous as trying to flap man-made wings, but many adventuresome people tried it in the late 1700s. In some areas, hot-air ballooning became a popular sport.

Soon afterward, large balloons were devised that were filled with hydrogen gas instead of hot air. Some of these balloons could lift large weights. The principles of floating on water and in air were discovered by Archimedes in about 300 B.C., but throughout the next 2,000 years, few people had tried to build something that would float in air.

Here is a technical illustration showing several early balloon designs.

During the 1900s, several commercial hydrogen-filled air ships were built. They could support a cabin that contained over a hundred people. These devices could fly people across long distances. They were on their way to becoming an important means of travel, until one of the air ships, known as the Hindenburg, crashed and burned in front of a large crowd in 1937 in New Jersey. People realized how unsafe they were and commercial travel by air ships soon came to an end.

During the late 1800s, gliders were developed as a result of careful observations about how birds glide through the air without flapping their wings. Once some controls were added, gliding became a popular sport in some areas. The pilot had to find a suitable place to jump from with flat land below the jump-off place. If all went well (a big IF), the glider would continue moving forward while slowly descending until it gently landed on the ground. Gliders worked on the same principles as paper and model airplanes. Before motorized airplanes could be built, models of gliders had to be designed and tested over and over.

The next big step in flight came in December of 1903 when Orville and Wilbur Wright attached a gasoline-powered motor to a glider and made the first powered flight in written history. They had spent months designing, testing, building, and rebuilding their plane. Their first flight was ignored by most people who heard about it, but a few years later, the brothers had greatly improved their designs. The brothers sold patents to the military and other businesses. Inventors from all over Europe and the United States went to work on their own inventions to create a flying machine. Once the basic designs were in place, progress in airplane flight was rapid.

During the 1930s and 1940s, rocket power was greatly developed. Rocket propulsion had been discovered centuries earlier, but there was almost no technology developed from it in Europe and America. Then during World War II, German scientists designed and built powerful rockets that were used to carry explosive bombs and attack Allied countries. After the war, this same technology was used by Russia and the United States to build rockets that would take men into space, and eventually to the moon.

36

Making Connections

Sometimes new technologies can only be made to work when two or more inventions are combined. That was true of airplanes. First, gliders were developed that had controls that could cause the plane to turn right and left and to rise and fall as needed. Then big coal-powered steam engines were modified and lightweight gasoline-powered engines were developed. In order to be safe and last for a long time, the airplanes had to be made from lightweight, durable materials, mostly metals, which were also being developed. During the late 1800s and the early 1900s, all the necessary discoveries were in place.

Dig Deeper

Rip Van Winkle is a fictional story about a man who went to sleep and slept for many years before he woke up. Write a similar story about someone from your hometown who went to sleep in 1900 and woke up 70 years later in 1970.

Interview an older person you know about the first time they heard about and saw an airplane or some kind of hot-air or hydrogen-filled blimp. You could also ask about what they remember their parents telling them. Write both your questions and the responses.

What Did You Learn?

1. In what year did Orville and Wilbur Wright make their first flight on a glider that was powered by a gasoline motor?
2. What did the Wright brothers have to spend months doing before they were able to test their first airplane?
3. Gliding was a popular hobby in the 1800s. Explain how a person could use a glider and travel hundreds of feet through the air.
4. Who discovered the principle of floating in air? In about what year was this discovery made?
5. Another popular hobby that began in the late 1700s was hot-air ballooning. What chemical eventually replaced the hot air and allowed people to travel long distances through the air?
6. Commercial air ships were beginning to be used in the early 1900s to carry passengers long distances through the air. What famous air ship burned and crashed, putting an end to travel by air ships?
7. Why were people never able to attach artificial wings to their bodies and fly?
8. Could an airplane be built that would take a man to the moon? What kind of device was used to transport men to the moon?
9. List 3 inventions that had to be developed before someone could build a motor-powered airplane.

The first successful flight was with the Wright Flyer, by the Wright brothers. The Flyer traveled 120 feet in 12 seconds at 10:35 a.m. at Kill Devil Hills, North Carolina.

37

What Did You Learn?

1. In what year did Orville and Wilbur Wright make their first flight on a glider that was powered by a gasoline motor? *1903*
2. What did the Wright brothers have to spend months doing before they were able to actually fly the first airplane? *Designing, testing, building, and rebuilding their plane.*
3. Gliding was a popular hobby in the 1800s. Explain how a person could use a glider and travel hundreds of feet through the air. *The pilot had to find a suitable place to jump from with flat land below the jump-off place. If all went well, the glider would continue moving forward while slowly descending until it gently landed on the ground. Gliders worked on the same principles as paper and model airplanes.*
4. Who discovered the principle of floating in air? In about what year was this discovery made? *Archimedes in about 300 B.C.*
5. Another popular hobby that began in the late 1700s was hot-air ballooning. What was the first gas to replace the hot air that allowed people to travel long distances through the air? *Hydrogen*
6. Commercial airships were beginning to be used in the early 1900s to carry passengers long distances through the air. What famous airship burned and crashed, putting an end to travel by air ships? *The Hindenburg*
7. Why were people never able to attach artificial wings to their bodies and fly? *People were too heavy and their muscles were too weak to fly.*
8. Could an airplane be built that would take a man to the moon? What kind of device was used to transport men to the moon? *Airplanes can only fly through air. They cannot fly through space. Rockets were used to take men to the moon and back.*
9. List 3 inventions that had to be developed before someone could build a motor-powered airplane. *First, gliders had to be developed that had controls that could cause the plane to turn right and left and to rise and fall as needed. Then big coal-powered steam engines had to be modified into light-weight gasoline-powered engines. Light-weight, durable materials, mostly metals, had to be developed which were used to make the airplanes.*

INVESTIGATION #9

Rockets and Space Exploration

Think about This Rockets were used primarily as a way to carry destructive weapons during World War II. Since then, rockets have been used to carry spacecraft and satellites into space. Rockets have enabled scientists to explore the solar system, send communication satellites and weather satellites into orbit, and put a series of GPS (global positioning system) satellites into place.

There have been many ways in which rockets have been used in this century. However, we will focus in this lesson on how rockets work and how they are able to put satellites into space. Why do you think rockets had to be used to put satellites into space rather than airplanes or balloons?

This Delta rocket launched on June 21, 1975 at Cape Canaveral, Florida, held Goddard's 8th Orbiting Solar Observatory.

Procedure and Observations **(A)** Cut a 4" x 4" square piece of paper. Put a strip of tape along one side of the paper. Let the tape overhang the paper about ¼". Then put another strip of tape along an adjacent side of the paper. **(B)** Remove the lid from the canister and tape the paper about ¼" from the open end. **(C)** The tube will be a little longer than the canister.

A B C

The Investigative Problems

How do rockets work? What new frontiers did rockets open up that went beyond airplanes?

38

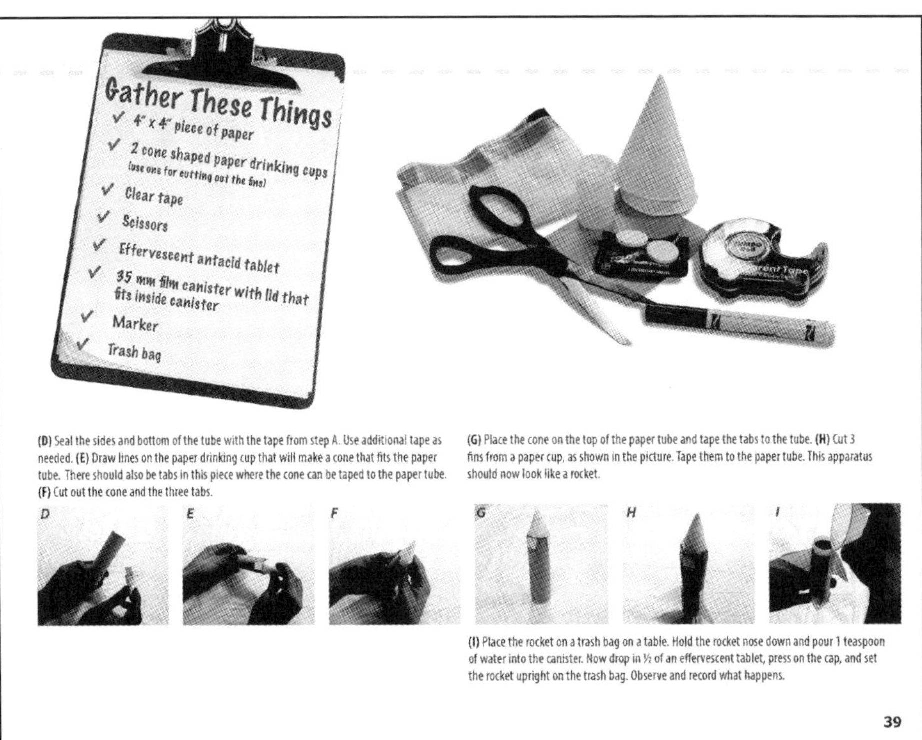

Gather These Things

- ✓ 4" x 4" piece of paper
- ✓ 2 cone shaped paper drinking cups (use one for cutting out the fins)
- ✓ Clear tape
- ✓ Scissors
- ✓ Effervescent antacid tablet
- ✓ 35 mm film canister with lid that fits inside canister
- ✓ Marker
- ✓ Trash bag

(D) Seal the sides and bottom of the tube with the tape from step A. Use additional tape as needed. **(E)** Draw lines on the paper drinking cup that will make a cone that fits the paper tube. There should also be tabs in this piece where the cone can be taped to the paper tube. **(F)** Cut out the cone and the three tabs.

(G) Place the cone on the top of the paper tube and tape the tabs to the tube. **(H)** Cut 3 fins from a paper cup, as shown in the picture. Tape them to the paper tube. This apparatus should now look like a rocket.

D E F G H I

(I) Place the rocket on a trash bag on a table. Hold the rocket nose down and pour 1 teaspoon of water into the canister. Now drop in ½ of an effervescent tablet, press on the cap, and set the rocket upright on the trash bag. Observe and record what happens.

39

OBJECTIVES

1. Rockets and airplanes work in different ways. Rockets can fly though both air and space, but airplanes can only fly through air.
2. Rockets operate on the basis of Newton's third law of motion that says "For every action, there is an equal and opposite reaction."
3. Rockets can achieve very fast speeds in space because there is virtually no air to exert friction on them.
4. Rockets have provided great benefits to mankind and they have also created great dangers.
5. Wehner von Braun was a rocket scientist who led a team that developed the most powerful rockets in the world. They were used during World War II to carry bombs to England, but after the war von Braun and his staff surrendered to American troops and came to America, where he eventually became head of NASA and built the rockets that sent men to the moon. He became a committed Christian and a creationist.

NOTE

If students have trouble finding the correct type of film canister or other items in the supply list, this is an alternative example of how rockets work according to Newton's third law of motion.

Tape a large straw to the side of a small paper sack. Thread a fishing line or other heavy string through the straw and tie both ends to items that are secure. The string should be at least three meters long. Blow up a balloon and put it inside the paper sack. Don't tie the end of the balloon. Release the balloon and observe what happens.

The physics of rockets was explained more fully in the book *Forces and Motion,* published by Master Books and authored by Tom DeRosa and Carolyn Reeves. You might find it helpful to review some of these principles with the students.

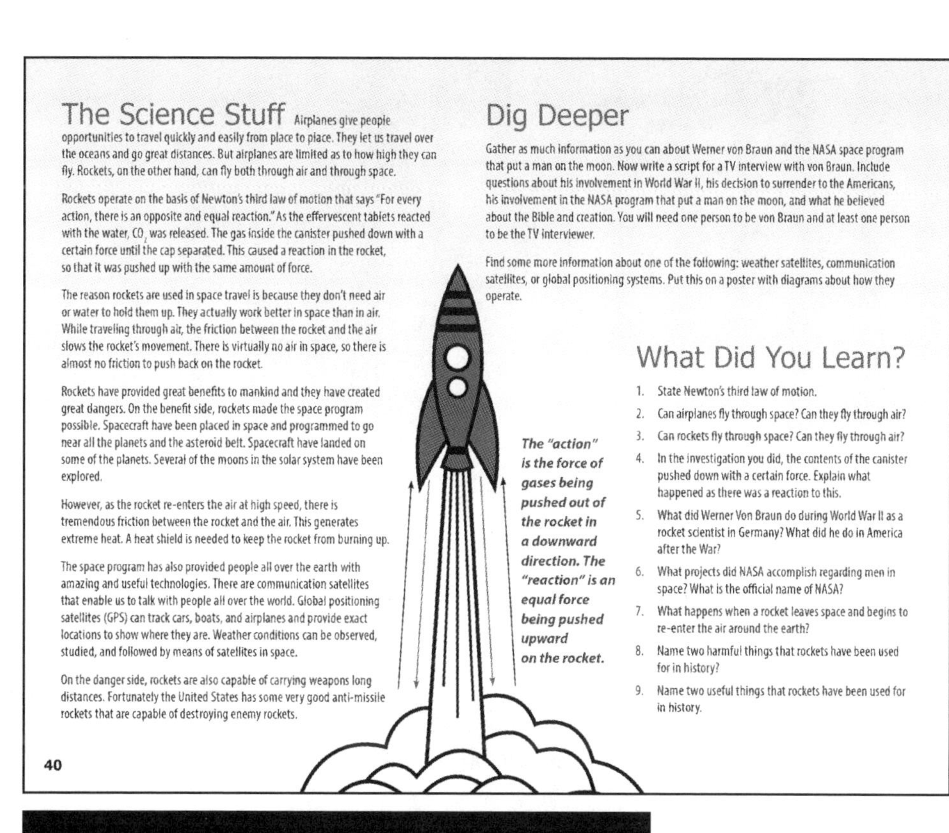

The Science Stuff

Airplanes give people opportunities to travel quickly and easily from place to place. They let us travel over the oceans and go great distances. But airplanes are limited as to how high they can fly. Rockets, on the other hand, can fly both through air and through space.

Rockets operate on the basis of Newton's third law of motion that says "For every action, there is an opposite and equal reaction." As the effervescent tablets reacted with the water, CO_2 was released. The gas inside the canister pushed down with a certain force until the cap separated. This caused a reaction in the rocket, so that it was pushed up with the same amount of force.

The reason rockets are used in space travel is because they don't need air or water to hold them up. They actually work better in space than in air. While traveling through air, the friction between the rocket and the air slows the rocket's movement. There is virtually no air in space, so there is almost no friction to push back on the rocket.

Rockets have provided great benefits to mankind and they have created great dangers. On the benefit side, rockets made the space program possible. Spacecraft have been placed in space and programmed to go near all the planets and the asteroid belt. Spacecraft have landed on some of the planets. Several of the moons in the solar system have been explored.

However, as the rocket re-enters the air at high speed, there is tremendous friction between the rocket and the air. This generates extreme heat. A heat shield is needed to keep the rocket from burning up.

The space program has also provided people all over the earth with amazing and useful technologies. There are communication satellites that enable us to talk with people all over the world. Global positioning satellites (GPS) can track cars, boats, and airplanes and provide exact locations to show where they are. Weather conditions can be observed, studied, and followed by means of satellites in space.

On the danger side, rockets are also capable of carrying weapons long distances. Fortunately the United States has some very good anti-missile rockets that are capable of destroying enemy rockets.

Dig Deeper

Gather as much information as you can about Werner von Braun and the NASA space program that put a man on the moon. Now write a script for a TV interview with von Braun. Include questions about his involvement in World War II, his decision to surrender to the Americans, his involvement in the NASA program that put a man on the moon, and what he believed about the Bible and creation. You will need one person to be von Braun and at least one person to be the TV interviewer.

Find some more information about one of the following: weather satellites, communication satellites, or global positioning systems. Put this on a poster with diagrams about how they operate.

The "action" is the force of gases being pushed out of the rocket in a downward direction. The "reaction" is an equal force being pushed upward on the rocket.

What Did You Learn?

1. State Newton's third law of motion.
2. Can airplanes fly through space? Can they fly through air?
3. Can rockets fly through space? Can they fly through air?
4. In the investigation you did, the contents of the canister pushed down with a certain force. Explain what happened as there was a reaction to this.
5. What did Werner Von Braun do during World War II as a rocket scientist in Germany? What did he do in America after the War?
6. What projects did NASA accomplish regarding men in space? What is the official name of NASA?
7. What happens when a rocket leaves space and begins to re-enter the air around the earth?
8. Name two harmful things that rockets have been used for in history?
9. Name two useful things that rockets have been used for in history.

40

Werner Von Braun
1912–1977

Making Connections

Werner Von Braun was a brilliant young rocket scientist in Germany when World War II began. He was given the job of building rockets that could travel for long distances. These rockets were used to carry bombs that were dropped on London and other cities in England, because England refused to surrender to Germany. In 1945, the Germans and their partners were finally defeated by the other European countries, Russia, and America. As the war was ending, Von Braun and a number of other German rocket scientists, technicians, and their families stayed in hiding until they found some American troops. They immediately surrendered to the first American troops they could find, because they did not want to be forced to go to Russia or to remain in Germany. They moved to the United States and worked to build rockets here. After several years, Von Braun became head of NASA (National Aeronautics and Space Administration), the group that built the rockets that carried men to the moon. Von Braun was known to his fellow workers in America as a committed Christian who spoke openly about his belief in a Creator.

41

What Did You Learn?

1. State Newton's third law of motion. *For every push, there is an equal push back. For every pull, there is an equal pull back.*

2. Can airplanes fly through space? Can they fly through air? *Airplanes cannot fly through space, but they can fly through air.*

3. Can rockets fly through space? Can they fly through air? *Rockets can fly through space, and they can also fly through air.*

4. In the investigation you did, the contents of the canister were pushed down with a certain force. Use Newton's third law of motion to explain what happened as there was a reaction to this. *When the contents of the canister were pushed down, the contents pushed back on the canister, so that it was pushed up with the same amount of force.*

5. What did Werner Von Braun do during World War II as a rocket scientist in Germany? What did he do in America after the war? *He helped design and build rockets for the Germans during World War II. After the war he came to the United States and became the head of the team that designed and built rockets that put Americans on the moon.*

6. What projects did NASA accomplish regarding men in space? What is the official name of NASA? *One of NASA's projects was to carry men into space and to the moon on rockets. National Aeronautics and Space Administration*

7. What happens when a spacecraft leaves space and begins to re-enter the air around the earth? *The spacecraft is traveling at a high rate of speed when it re-enters the earth's atmosphere. The friction between the air and the spacecraft creates tremendous heat. The spacecraft must be protected by a heat shield to keep it from burning up during re-entry.*

8. Name a harmful way rockets have been used in history. *They have been used to carry bombs and other weapons long distances.*

9. Name two or more useful ways in which rockets have been used in history. *Rockets made the space exploration possible. The space program has also provided people all over the earth with useful technologies, such as communication satellites that enable us to talk with people all over the world; global positioning satellites (GPS) that can track cars, boats, and airplanes and provide exact locations to show where they are; and weather satellites that can observe and study weather conditions.*

Investigation #10

The Earth in Space

Think about This Our little earth may seem insignificant compared to the vastness of the universe, but in God's sight, earth is a special place. Our Creator made all the right conditions necessary for life. His crowning creation was humans, who were made in His own image.

All the conditions for maintaining life are found on the earth. Even more amazing is that the size, density, and positions of the sun, moon, and other planets play a major role in making the earth just right for life. The position of the earth in our solar system is precisely where it should be in order for life to survive and flourish. If we were too close to the sun, water would become a vapor and life would probably not be possible. If we were too far away from the sun, the earth would be a ball of ice. The earth has the right amount of gravitational attraction to keep an atmosphere, without which there would be no liquid water on earth. Do you think life on earth could exist if conditions were very different than what they are now?

The Investigative Problems

What are the things that make life on earth possible? What would happen if these things changed?

42

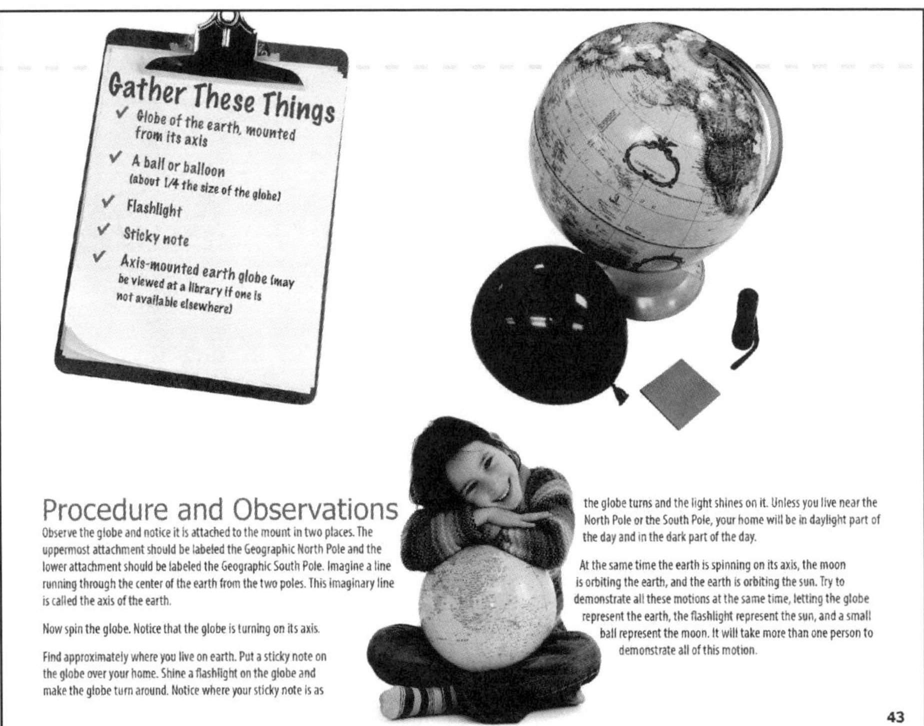

Gather These Things

- ✓ Globe of the earth, mounted from its axis
- ✓ A ball or balloon (about 1/4 the size of the globe)
- ✓ Flashlight
- ✓ Sticky note
- ✓ Axis-mounted earth globe (may be viewed at a library if one is not available elsewhere)

Procedure and Observations

Observe the globe and notice it is attached to the mount in two places. The uppermost attachment should be labeled the Geographic North Pole and the lower attachment should be labeled the Geographic South Pole. Imagine a line running through the center of the earth from the two poles. This imaginary line is called the axis of the earth.

Now spin the globe. Notice that the globe is turning on its axis.

Find approximately where you live on earth. Put a sticky note on the globe over your home. Shine a flashlight on the globe and make the globe turn around. Notice where your sticky note is as the globe turns and the light shines on it. Unless you live near the North Pole or the South Pole, your home will be in daylight part of the day and in the dark part of the day.

At the same time the earth is spinning on its axis, the moon is orbiting the earth, and the earth is orbiting the sun. Try to demonstrate all these motions at the same time, letting the globe represent the earth, the flashlight represent the sun, and a small ball represent the moon. It will take more than one person to demonstrate all of this motion.

43

Objectives

1. There are many conditions on the earth that make it a suitable place for man and other living things to live.
2. The masses and the distances of the sun and the moon also make life possible.
3. The gravitational pull of the large gas planets in our solar system helps to protect the earth from being hit by meteorites.
4. The sun provides heat and light to the earth, which are essential for life. Green plants are able to use sunlight during the process of photosynthesis and make food from carbon dioxide and water. Energy from the sun is stored in the food made by plants.

Note

G. Gonzales wrote *The Privileged Planet* in which he details a number of conditions that must be just right in order for life to exist on the earth. Many in the science community were extremely critical of the book, because it supported the idea of a supernatural design. As Christians, we would expect all the conditions necessary for life to be present, down to the smallest details. Many evolutionary naturalists strenuously object to the idea that God created the earth for the purpose of being a place for man and living things to live.

What Did You Learn?

1. What is the axis of the earth? *The earth's axis is an imaginary line that connects the North Pole and the South Pole. The earth rotates on its axis.*
2. How many times does the earth spin on its axis during a 24-hour day? *It spins on its axis one time during a 24-hour day.*

The Science Stuff

Remember, the earth spins on its axis one time during a 24-hour day. The moon spins on its axis one time during a monthly lunar cycle. It takes about one month for the moon to make one orbit around the earth. It takes about 365 days for the earth to make one orbit around the sun.

It took astronomers hundreds of years to figure out how everything in our universe was moving. One of the earliest explanations for why the moon and stars appear to be moving was that the earth was stationary and the sun, moon, and stars traveled around the earth. This was logical, because it looks like the sun moves through the sky in day. It also looks like the moon and stars move through the sky at night.

We know today that the earth orbits the sun, and the moon orbits the earth. These movements are similar to the way you demonstrated their movements.

The movement doesn't end there. The sun also moves in the Milky Way Galaxy as the galaxy spins around. It also appears that the galaxies themselves are moving farther and farther away from each other.

Since space travel has begun, there is a greater awareness of what is necessary for living things to survive. It is difficult to know all the conditions that make the earth suitable for life. But we will briefly mention a few that are essential for life as we know it.

1 winter, 2 spring, 3 summer, 4 fall

Diagram of the sun and earth at different seasons in the Northern Hemisphere.

The tilt of the earth on its axis is what gives us different seasons every year, so that there is a time to plant, and a time to harvest food. The sun's rays are more direct, and therefore, hotter during summer months. The sun's rays are slanted, and therefore, colder during winter months. If the earth were like Venus, which is not tilted on its axis, we would not have spring, summer, fall, and winter seasons.

Green plants use sunlight during the process of photosynthesis and make food from carbon dioxide and water. This food is stored in green plants and contains stored energy from the sun in the form of chemical energy. Green plants also give off oxygen during this same process. The earth maintains a balance of carbon dioxide and oxygen at all times.

There is an abundance of water on earth. The perfectly balanced water cycle keeps the amount of rain and snow that falls on the earth coming at the same rate that water on the earth evaporates into the air. The earth is not too close or too far away from the sun to upset the water cycle.

The moon is just the right size and distance from the earth to cause tides and produce a tidal zone that is an important part of the earth's food supply. If the moon were closer or larger, its gravitational pull would cause huge tides. There would be violent flooding and no shallow-water communities to produce a needed part of the food chain. It is also interesting that the moon is exactly the right size to just cover the sun during an eclipse of the sun.

The earth's atmosphere is held in place by just the right amount of gravitational force. There is plenty of oxygen in the air to live, but not so much that fires would be uncontrollable. The air pressure around us balances internal pressures.

The outer gas giants function as a protective barrier against meteorites and comets that might come in from outer space. The gravitational pull of the outer gas giant planets attract incoming asteroids and space rocks and prevent them from crashing on the earth.

The earth's magnetic field acts as a shield for many kinds of dangerous particles and rays that bombard the earth from outer space.

There are many other conditions that make the earth "just right" for life to exist and flourish for thousands of years.

44

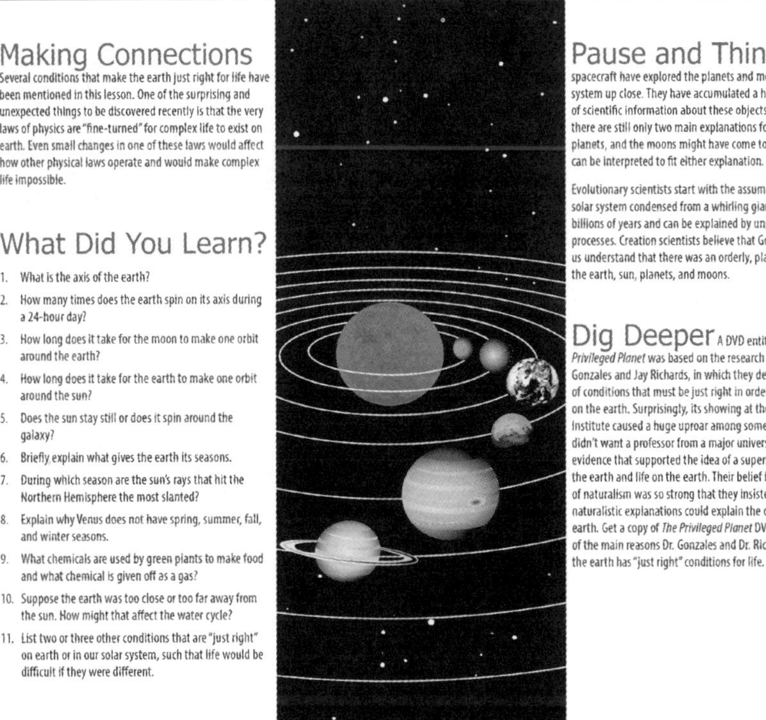

Making Connections

Several conditions that make the earth just right for life have been mentioned in this lesson. One of the surprising and unexpected things to be discovered recently is that the very laws of physics are "fine-turned" for complex life to exist on earth. Even small changes in one of these laws would affect how other physical laws operate and would make complex life impossible.

What Did You Learn?

1. What is the axis of the earth?
2. How many times does the earth spin on its axis during a 24-hour day?
3. How long does it take for the moon to make one orbit around the earth?
4. How long does it take for the earth to make one orbit around the sun?
5. Does the sun stay still or does it spin around the galaxy?
6. Briefly explain what gives the earth its seasons.
7. During which season are the sun's rays that hit the Northern Hemisphere the most slanted?
8. Explain why Venus does not have spring, summer, fall, and winter seasons.
9. What chemicals are used by green plants to make food and what chemical is given off as a gas?
10. Suppose the earth was too close or too far away from the sun. How might that affect the water cycle?
11. List two or three other conditions that are "just right" on earth or in our solar system, such that life would be difficult if they were different.

Pause and Think

Numerous spacecraft have explored the planets and moons in the solar system up close. They have accumulated a huge amount of scientific information about these objects. However, there are still only two main explanations for how the sun, planets, and the moons might have come to exist. The facts can be interpreted to fit either explanation.

Evolutionary scientists start with the assumption that the solar system condensed from a whirling giant nebula over billions of years and can be explained by unplanned natural processes. Creation scientists believe that God's Word helps us understand that there was an orderly, planned creation of the earth, sun, planets, and moons.

Dig Deeper

A DVD entitled *The Privileged Planet* was based on the research of Guillermo Gonzales and Jay Richards, in which they detail a number of conditions that must be just right in order for life to exist on the earth. Surprisingly, its showing at the Smithsonian Institute caused a huge uproar among some scientists who didn't want a professor from a major university to show evidence that supported the idea of a supernatural design of the earth and life on the earth. Their belief in the philosophy of naturalism was so strong that they insisted that only naturalistic explanations could explain the origin of the earth. Get a copy of *The Privileged Planet* DVD and make a list of the main reasons Dr. Gonzales and Dr. Richards gave that the earth has "just right" conditions for life.

45

3. How long does it take for the moon to make one orbit around the earth? *About one month.*
4. How long does it take for the earth to make one orbit around the sun? *It takes about 365 days or one year.*
5. Does the sun stay still or does it move around the galaxy? *The sun moves around the galaxy.*
6. Briefly explain what gives the earth its seasons. *The tilt of the earth on its axis is what gives us different seasons every year. The sun's rays are more direct, and therefore hotter during summer months. The sun's rays are slanted, and therefore colder during winter months.*
7. During which season are the sun's rays that hit the Northern Hemisphere the most slanted? *During winter.*
8. Explain why Venus does not have spring, summer, fall, and winter seasons. *It is not tilted on its axis.*
9. What chemicals are used by green plants to make food and what chemical is given off as a gas? *Green plants use carbon dioxide and water to make food. They give off oxygen as a gas.*
10. Suppose the earth was too close or too far away from the sun. How might that affect the water cycle? *The perfectly balanced water cycle keeps the amount of rain and snow that falls on the earth coming at the same rate that water on the earth evaporates into the air. The earth is not too close or too far away from the sun to upset the water cycle.*
11. List two or three other conditions that are "just right" on earth or in our solar system, such that life would be difficult if they were different. *There is the right amount of gravitational force to hold the earth's atmosphere in place. There is enough oxygen to live, but not so much that fires would be uncontrollable. Air pressure is enough to balance internal pressures. The earth's magnetic field acts as a shield for many kinds of dangerous particles and rays that bombard the earth from outer space.*

Objectives

1. Without an atmosphere, the earth would look much like the moon and would be equally as barren.

2. The earth's atmosphere is held in place by a sufficient amount of gravitational attraction to the earth. The moon's gravitational pull is not sufficient to maintain an atmosphere.

3. The earth's atmosphere "traps" heat from the sun and prevents it from escaping into space. It prevents the water on earth from boiling away. It causes air pressure to exist around everything on earth and is the right amount to balance the internal pressures of living things.

4. Oxygen in the atmosphere causes most meteorites that come close to the earth to burn up before they hit the earth. Combined with the earth's magnetosphere, harmful rays and particles from the sun and space are prevented from striking the earth in large numbers.

5. The earth's atmosphere is made up of about 20 percent oxygen and about 80 percent nitrogen. This is the right amount for breathing. Too much oxygen in the air would cause fires to burn out of control. The amount of oxygen in the air stays at a constant amount. Plants produce oxygen during photosynthesis at the same rate animals use oxygen for respiration.

6. Weather changes occur because the earth has an atmosphere that varies in moisture content and in temperature.

7. The layers of the atmosphere in order are troposphere, stratosphere, mesosphere, thermosphere, ionosphere.

INVESTIGATION #11

The Earth's Atmosphere

Think about This Katlin read that people might one day live in colonies on the moon. She was intrigued by the possibility of making a visit to the moon and seeing the whole earth in the sky. She wanted to see the moon's black sky with a clear view of all the stars.

The book continued to explain that, without an atmosphere, there would be many problems in maintaining such a colony. For example, no one could take a walk on the moon and survive unless they were wearing a special pressurized space suit. They would have to live in pressurized buildings where all the air was recycled. If meteoroids were encountered from space, they wouldn't burn up before hitting the moon and might hit or destroy their space station. There would be no rain, clouds, winds, trees, or rivers. The days would be extremely hot and the nights extremely cold.

Did you know that the earth would be a lot like the moon if the earth had no atmosphere? Would you like to spend some time in a colony on the moon?

The Investigative Problems

What would the earth be like if there was no atmosphere around it? How does the earth's magnetosphere help protect the earth?

46

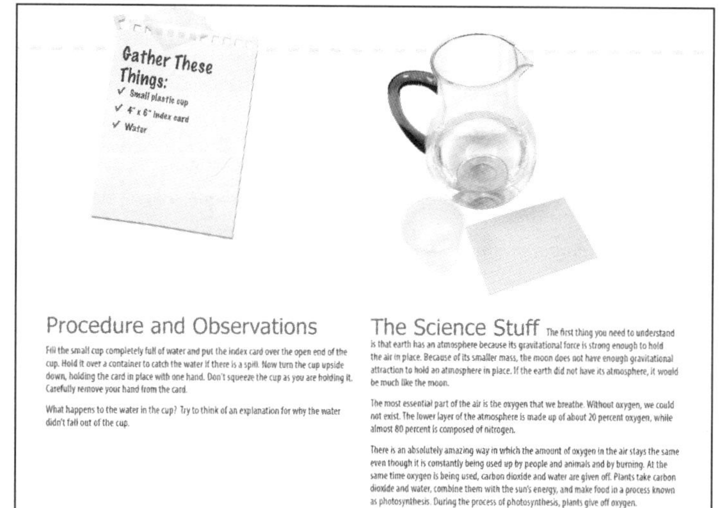

Gather These Things:

- ✓ Small plastic cup
- ✓ 4" x 6" index card
- ✓ Water

Procedure and Observations

Fill the small cup completely full of water and put the index card over the open end of the cup. Hold it over a container to catch the water if there is a spill. Now turn the cup upside down, holding the card in place with one hand. Don't squeeze the cup as you are holding it. Carefully remove your hand from the card.

What happens to the water in the cup? Try to think of an explanation for why the water didn't fall out of the cup.

The Science Stuff The first thing you need to understand is that earth has an atmosphere because its gravitational force is strong enough to hold the air in place. Because of its smaller mass, the moon does not have enough gravitational attraction to hold an atmosphere in place. If the earth did not have its atmosphere, it would be much like the moon.

The most essential part of the air is the oxygen that we breathe. Without oxygen, we could not exist. The lower layer of the atmosphere is made up of about 20 percent oxygen, while almost 80 percent is composed of nitrogen.

There is an absolutely amazing way in which the amount of oxygen in the air stays the same even though it is constantly being used up by people and animals and by burning. At the same time oxygen is being used, carbon dioxide and water are given off. Plants take carbon dioxide and water, combine them with the sun's energy, and make food in a process known as photosynthesis. During the process of photosynthesis, plants give off oxygen.

47

Note

This may be a good opportunity to discuss some of the problems of air pollution in a balanced way. We suggest that you refer to some of the articles about climate change and the atmosphere in the recommended reading sites to find a good balance.

What Did You Learn?

1. How do plants and animals maintain a balance of the amount of carbon dioxide and oxygen in the air? *Animals breathe out carbon dioxide and water, which plants use to make food. Plants give off oxygen which animals breathe in.*

2. How much of the air we breathe is made up of oxygen? *The lower layer of the atmosphere is made up of about 20 percent oxygen.*

3. What causes air pressure on the earth? What is the average pressure at sea level? *Air pressure is the result of the weight of the air above. It is about 14.7 pounds on every square inch at sea level.*

4. Could astronauts survive at the edge of the atmosphere without a pressurized cabin or suit? *No, they would quickly die.*

5. Why do planets and moons with no atmosphere or very thin air not have liquid water on them? *When there is no air pressure, the water boils away, even in cold temperatures.*

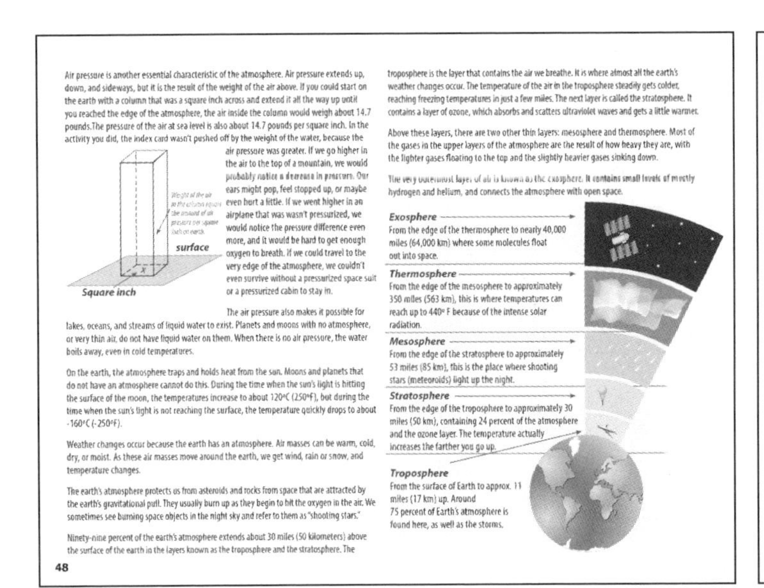

Air pressure is another essential characteristic of the atmosphere. Air pressure extends up, down, and sideways, but it is the result of the weight of the air above. If you could start on the earth with a column that was a square inch across and extend it all the way up until you reached the edge of the atmosphere, the air inside the column would weigh about 14.7 pounds.The pressure of the air at sea level is also about 14.7 pounds per square inch. In the activity you did, the index card wasn't pushed off by the weight of the water, because the air pressure was greater. If we go higher in the air to the top of a mountain, we would probably notice a decrease in pressure. Our ears might pop, feel stopped up, or maybe even hurt a little. If we went higher in an airplane that was wasn't pressurized, we would notice the pressure difference even more, and it would be hard to get enough oxygen to breath. If we could travel to the very edge of the atmosphere, we couldn't even survive without a pressurized space suit or a pressurized cabin to stay in.

The air pressure also makes it possible for lakes, oceans, and streams of liquid water to exist. Planets and moons with no atmosphere, or very thin air, do not have liquid water on them. When there is no air pressure, the water boils away, even in cold temperatures.

On the earth, the atmosphere traps and holds heat from the sun. Moons and planets that do not have an atmosphere cannot do this. During the time when the sun's light is hitting the surface of the moon, the temperatures increase to about 120°C (250°F), but during the time when the sun's light is not reaching the surface, the temperature quickly drops to about -160°C (-250°F).

Weather changes occur because the earth has an atmosphere. Air masses can be warm, cold, dry, or moist. As these air masses move around the earth, we get wind, rain or snow, and temperature changes.

The earth's atmosphere protects us from asteroids and rocks from space that are attracted by the earth's gravitational pull. They usually burn up as they begin to hit the oxygen in the air. We sometimes see burning space objects in the night sky and refer to them as "shooting stars."

Ninety-nine percent of the earth's atmosphere extends about 30 miles (50 kilometers) above the surface of the earth in the layers known as the troposphere and the stratosphere. The troposphere is the layer that contains the air we breathe. It is where almost all the earth's weather changes occur. The temperature of the air in the troposphere steadily gets colder, reaching freezing temperatures in just a few miles. The next layer is called the stratosphere. It contains a layer of ozone, which absorbs and scatters ultraviolet waves and gets a little warmer.

Above these layers, there are two other thin layers: mesosphere and thermosphere. Most of the gases in the upper layers of the atmosphere are the result of how heavy they are, with the lighter gases floating to the top and the slightly heavier gases sinking down.

The very outermost layer of air is known as the exosphere. It contains small levels of mostly hydrogen and helium, and connects the atmosphere with open space.

48

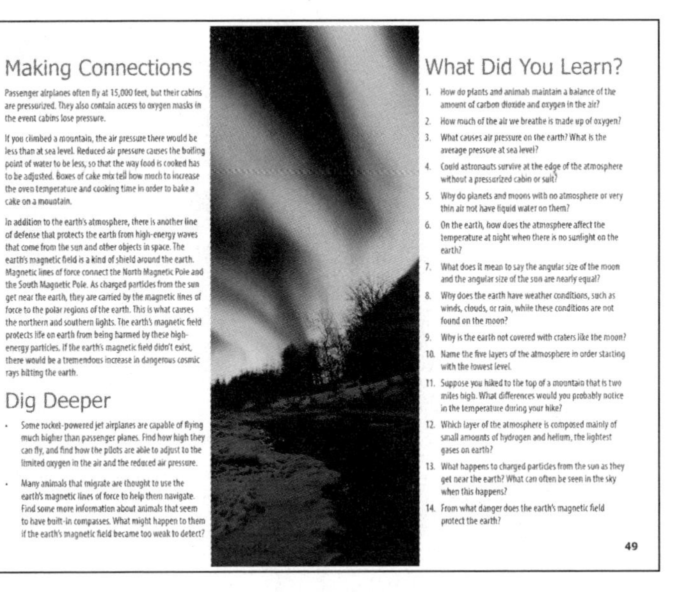

Making Connections

Passenger airplanes often fly at 15,000 feet, but their cabins are pressurized. They also contain access to oxygen masks in the event cabins lose pressure.

If you climbed a mountain, the air pressure there would be less than at sea level. Reduced air pressure causes the boiling point of water to be less, so that the way food is cooked has to be adjusted. Boxes of cake mix tell how much to increase the oven temperature and cooking time in order to bake a cake on a mountain.

In addition to the earth's atmosphere, there is another line of defense that protects the earth from high-energy waves that come from the sun and other objects in space. The earth's magnetic field is a kind of shield around the earth. Magnetic lines of force connect the North Magnetic Pole and the South Magnetic Pole. As charged particles from the sun get near the earth, they are carried by the magnetic lines of force to the polar regions of the earth. This is what causes the northern and southern lights. The earth's magnetic field protects life on earth from being harmed by these high-energy particles. If the earth's magnetic field didn't exist, there would be a tremendous increase in dangerous cosmic rays hitting the earth.

Dig Deeper

- Some rocket-powered jet airplanes are capable of flying much higher than passenger planes. Find how high they can fly, and find how the pilots are able to adjust to the limited oxygen in the air and the reduced air pressure.
- Many animals that migrate are thought to use the earth's magnetic lines of force to help them navigate. Find some more information about animals that seem to have built-in compasses. What might happen to them if the earth's magnetic field became too weak to detect?

What Did You Learn?

1. How do plants and animals maintain a balance of the amount of carbon dioxide and oxygen in the air?
2. How much of the air we breathe is made up of oxygen?
3. What causes air pressure on the earth? What is the average pressure at sea level?
4. Could astronauts survive at the edge of the atmosphere without a pressurized cabin or suit?
5. Why do planets and moons with no atmosphere or very thin air not have liquid water on them?
6. On the earth, how does the atmosphere affect the temperature at night when there is no sunlight on the earth?
7. What does it mean to say the angular size of the moon and the angular size of the sun are nearly equal?
8. Why does the earth have weather conditions, such as winds, clouds, or rain, while these conditions are not found on the moon?
9. Why is the earth not covered with craters like the moon?
10. Name the five layers of the atmosphere in order starting with the lowest level.
11. Suppose you hiked to the top of a mountain that is two miles high. What differences would you probably notice in the temperature during your hike?
12. Which layer of the atmosphere is composed mainly of small amounts of hydrogen and helium, the lightest gases on earth?
13. What happens to charged particles from the sun as they get near the earth? What can often be seen in the sky when this happens?
14. From what danger does the earth's magnetic field protect the earth?

49

6. On the earth, how does the atmosphere affect the temperature at night when there is no sunlight on the earth? *On the earth, the atmosphere traps and holds heat from the sun.*

7. Briefly explain how the moon's temperature is different from the earth as a result of not having an atmosphere. *There is no atmosphere to trap and hold heat from the sun. When the sun's light is hitting the surface of the moon, the temperatures climb to about 250ºF, but when the sun's light is not reaching the surface, the temperature quickly drops to about -250ºF.*

8. Why does the earth have weather conditions, such as winds, clouds, or rain, while these conditions are not found on the moon? *Winds, clouds, and rain occur on the earth because the earth has an atmosphere. These conditions do not occur on the moon, because the moon does not have an atmosphere.*

9. Why is the earth not covered with craters like the moon? *Although asteroids and rocks from space are often attracted to the earth by gravity, they usually burn up as they begin to hit the oxygen in the air.*

10. Name the five layers of the atmosphere in order, starting with the lowest level. *Troposphere, stratosphere, mesosphere, thermosphere, and exosphere.*

11. Suppose you hiked to the top of a mountain that is two miles high. What differences would you probably notice in the temperature during your hike? *The temperature at the top of the mountain would probably be much colder than the temperature at the bottom of the mountain.*

12. Which layer of the atmosphere is composed mainly of small amounts of hydrogen and helium, the lightest gases on earth? *Exosphere*

13. What happens to charged particles from the sun as they get near the earth? What can often be seen in the sky when this happens? *As charged particles from the sun get near the earth, they are carried by the earth's magnetic lines of force to the polar regions of the earth. The northern and southern lights can often be seen in the sky when this happens by people who live close enough to the poles.*

14. From what danger does the earth's magnetic field protect the earth? *The earth's magnetic field protects life on earth from being harmed by high energy particles and cosmic rays.*

Objectives

1. The moon orbits the earth while the earth orbits the sun.
2. The moon makes a complete orbit around the earth about every month. During this same time, it make a complete rotation on its axis. This means that one side of the moon is never visible from the earth.
3. The size of the moon is exactly the right size to just cover the sun during a total eclipse of the sun.
4. The earth's gravitational pull keeps the moon in its orbital path around the earth. The sun and moon's gravitational pull causes the ocean tides.
5. There is no air around the moon. Because of this, there are no bodies of liquid water on the moon, there is no wind on the moon, and meteorites don't burn up before they hit the moon.
6. American astronauts were the first men to land on the moon. This occurred in 1969.
7. The moon is receding from the earth at the rate of about two inches per year.

INVESTIGATION #12

The Moon

Think about This The moment had finally come for Neil, Buzz, and Michael to complete their mission. They had planned and practiced for this for months, but there were still many unknowns and much danger in what they were about to do. The risky maneuver involved Neil and Buzz landing their craft on the ground below while Michael remained in another larger craft that continued moving in a pattern above the ground.

The landing of the two-man craft was controlled by rocket power. They were connected to headquarters by a sophisticated communication system including live TV transmission. Neil and Buzz began their descent and headed for the target area. Their altitude was called out frequently — 4,200 feet, 2,000 feet, 1,600 feet, but landmarks were slightly off, meaning they might land miles from the targeted landing. Then unexpected program alarms began sounding. Headquarters made the critical decision to continue on.

Realizing they had to avoid a field ahead containing several large boulders, Neil switched to manual control as Buzz called out altitude and speeds. They had to find a flat place to land, and soon. Elevation was 4 feet and dust was flying all around. Thirty seconds of fuel remained as Neil searched for a safer place to land. The people at headquarters waited in absolute silence to hear if they had made a safe landing. They burst into cheers when they finally heard Neil report in, "Tranquility Base here, Houston. The Eagle has landed."

Did you realize this was an account of when the first men landed on the moon on July 20, 1969, and that it was witnessed on TV by half a billion people around the world? What do you know about the Apollo 11 mission that carried Neil Armstrong, Buzz Aldrin, and Michael Collins to the moon and returned them safely to the earth?

The Investigative Problems

Why are there so many craters on the surface of the moon? What might happen if the moon's size and distance from the earth were different?

50

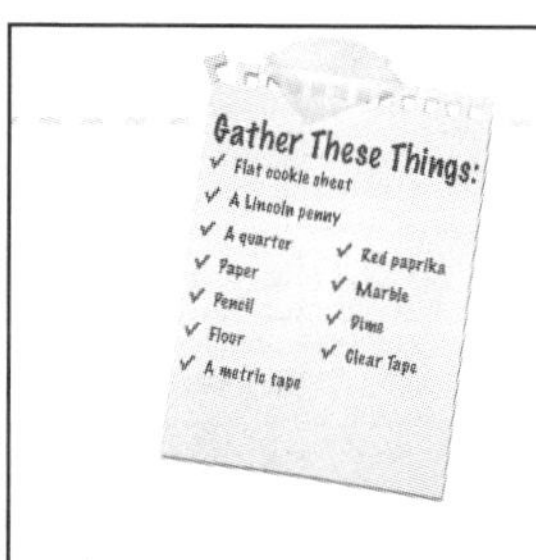

Procedure and Observations

Part A

Put a quarter flat on the table to represent earth. Place a penny flat on the table with the "heads" side facing up to represent the moon. Slide the penny around the quarter in a circle, like it was completing an orbit.

Lincoln's face on the penny should keep looking at you throughout its orbit. Make four drawings of the penny and the quarter. Show when the penny is ¼, ½, ¾, and all the way around the quarter. Is Lincoln's face turned toward the quarter in each of these drawings?

Repeat this exercise, but this time make sure Lincoln's face on the penny (the moon) always looks at the quarter (the earth) throughout its orbit. Make four drawings of the penny and the quarter. Show when the penny is ¼, ½, ¾, and all the way around the quarter. Is Lincoln's face always turned toward the quarter in each of these drawings?

Part B

This activity will demonstrate the effects of meteoroids landing on the moon. Take a large flat cookie sheet, fill it almost to the rim with white flour, then cover the surface of the flour with the red paprika. Stand above the cookie sheet, and drop marbles of various sizes onto the surface. Carefully pick up the marbles after they are dropped, trying not to disturb the flour. Observe the craters and rims that form from the impact. Try to have some impacts that overlay previous impacts. Repeat as needed.

Part C

This investigation will need to be done when there is a full moon that is visible. Measure the diameter of a dime with a metric ruler. Use clear tape to stick the dime to a window where a full moon can be seen. View the moon with one eye so that the dime just blocks out the full moon. Move closer or farther away from the dime until you find this position. Get a partner to help you use a metric tape (not a stiff ruler) to find the distance from beside your eye to the dime. Record this information. Divide the diameter of the coin into the distance from the coin to the eye and record your answer.

Hint: If all your measurements were accurate, you should have found that the diameter of the dime is about 18 mm and your calculated answer is about 110. If you repeat this activity with a larger coin, you will still get the calculated answer of about 110. This will work with a larger coin, because the distance from your eye to the coin will have to increase to just block out the full moon.

51

Note

Shining a light on two balls representing the earth and the moon is a valuable way of illustrating the phases of the moon. You might consider doing this as a demonstration. A flashlight would represent the sun; a larger ball, the earth; and a smaller ball, the moon. You would need to darken the room and have two students hold the balls. The moon would move around the earth while the "sun's light" remained on. Students have to understand that the earth is actually moving around the sun, but the sun will remain stationary for the demonstration. They have to understand that when it is night on one side of the earth, the people there can see only the part of the moon that is illuminated by the "sun." The moon's orbit is also tilted somewhat. All of these conditions enable people on earth to see the moon throughout the month.

This is an alternative activity to illustrate why we only see one side of the moon from the earth. Have one person stand and hold a sign that says "Earth." Have another person hold a sign that says "Moon." The moon should walk around the sun in two ways. (1) The first time, the moon student should face the same wall all the ways around the "Earth" by walking forward halfway and backward halfway. (2) The next time, the moon student should walk in such a way the moon always faces the "Earth" by walking sideways to maintain the same orientation throughout the orbit.

"Lunar recession" is a significant argument that the moon is not five billion years old. Have students multiply two inches by 1,000,000,000 years. Then divide that number by 63,360 inches (the number of inches in a mile) to see (hypothetically) how much closer to the earth the moon would have been just a billion years ago. A billion years ago, ocean tides would be enormous. If the moon was ever too close to the earth, the earth's gravitational pull would cause it to break up into pieces, probably becoming rings around the earth, like Saturn's rings.

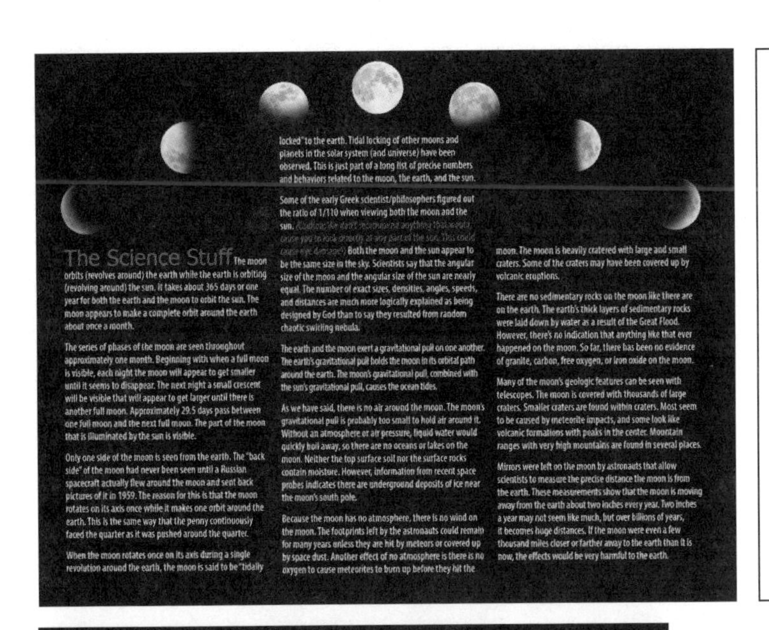

The Science Stuff

The moon orbits (revolves around) the earth while the earth is orbiting (revolving around) the sun. It takes about 365 days or one year for both the earth and the moon to orbit the sun. The moon appears to make a complete orbit around the earth about once a month.

The series of phases of the moon are seen throughout approximately one month. Beginning with when a full moon is visible, each night the moon will appear to get smaller until it seems to disappear. The next night a small crescent will be visible that will appear to get larger until there is another full moon. Approximately 29.5 days pass between one full moon and the next full moon. The part of the moon that is illuminated by the sun is visible.

Only one side of the moon is seen from the earth. The "back side" of the moon had never been seen until a Russian spacecraft actually flew around the moon and sent back pictures of it in 1959. The reason for this is that the moon rotates on its axis once while it makes one orbit around the earth. This is the same way that the penny continuously faced the quarter as it was pushed around the quarter.

When the moon rotates once on its axis during a single revolution around the earth, the moon is said to be "tidally locked" to the earth. Tidal locking of other moons and planets in the solar system (and universe) have been observed. This is just part of a long list of precise numbers and behaviors related to the moon, the earth, and the sun.

Some of the early Greek scientist/philosophers figured out the ratio of 1/110 when viewing both the moon and the sun. [illegible] Both the moon and the sun appear to be the same size in the sky. Scientists say that the angular size of the moon and the angular size of the sun are nearly equal. The number of exact sizes, densities, angles, speeds, and distances are much more logically explained as being designed by God than to say they resulted from random chaotic swirling nebula.

The earth and the moon exert a gravitational pull on one another. The earth's gravitational pull holds the moon in its orbital path around the earth. The moon's gravitational pull, combined with the sun's gravitational pull, causes the ocean tides.

As we have said, there is no air around the moon. The moon's gravitational pull is probably too small to hold air around it. Without an atmosphere or air pressure, liquid water would quickly boil away, so there are no oceans or lakes on the moon. Neither the top surface soil nor the surface rocks contain moisture. However, information from recent space probes indicates there are underground deposits of ice near the moon's south pole.

Because the moon has no atmosphere, there is no wind on the moon. The footprints left by the astronauts could remain for many years unless they are hit by meteors or covered up by space dust. Another effect of no atmosphere is there is no oxygen to cause meteorites to burn up before they hit the moon. The moon is heavily cratered with large and small craters. Some of the craters may have been covered up by volcanic eruptions.

There are no sedimentary rocks on the moon like there are on the earth. The earth's thick layers of sedimentary rocks were laid down by water as a result of the Great Flood. However, there's no indication that anything like that ever happened on the moon. So far, there has been no evidence of granite, carbon, free oxygen, or iron oxide on the moon.

Many of the moon's geologic features can be seen with telescopes. The moon is covered with thousands of large craters. Smaller craters are found within craters. Most seem to be caused by meteorite impacts, and some look like volcanic formations with peaks in the center. Mountain ranges with very high mountains are found in several places.

Mirrors were left on the moon by astronauts that allow scientists to measure the precise distance the moon is from the earth. These measurements show that the moon is moving away from the earth about two inches every year. Two inches a year may not seem like much, but over billions of years, it becomes huge distances. If the moon were even a few thousand miles closer or farther away to the earth than it is now, the effects would be very harmful to the earth.

Crater 302 on the moon's surface was photographed by the Apollo 10 astronauts in May of 1969. You can easily see the terraced walls of the crater and the cone in the center.

Making Connections

Some of the earth's earliest scientists used ratios, angles, and other geometric principles to estimate the distances from the earth to the moon and distances from the earth to the sun. They also estimated sizes of the earth, moon, and sun that were pretty accurate. We know Greeks were able to do this, but there are indications that this had also been done at a much earlier time.

The United States or some of the other countries may consider building colonies on the moon in the future. An area at the moon's South Pole has been explored and found to have a supply of water in the form of ice. This could be a source of drinking water and could also be a possible source of hydrogen for rocket fuel.

Dig Deeper

Do an Internet search for videos of the Apollo 11 landing and mission. Try to find one that shows what people around the world saw on TV. (Avoid confusion by not choosing the ones about aliens and hoaxes.) Describe their landing in your own words.

Make a chart comparing the similarities and differences between the moon and the earth.

Make a calendar for one month and leave enough room to draw the shape of the moon as seen every night. Keep a record of the moon phases every night for a month. Draw the moon and name the main phases as they appear. If there are some nights too cloudy for the moon to be seen, leave those dates blank. Include as many drawings as you can. Be sure the dates are correctly matched with the drawings of the moon. Try to predict the shape of the moon on any days you left blank.

What Did You Learn?

1. Does the moon orbit the earth while the earth orbits the sun?
2. How long does it take the earth to make one complete trip around the sun? About how long does it take the moon to make one complete trip around the earth?
3. Both the earth and the moon rotate on their axis. How long does it take for the earth to make one complete rotation? About how long does it take for the moon to make one complete rotation?
4. Why is only one side of the moon seen from the earth?
5. Sometimes only part of the moon is visible in the night sky from the earth. What determines which part of the moon is visible at night?
6. How does the earth's gravitational pull affect the moon? How does the moon's gravitational pull affect the earth?
7. In the investigation you did with the dime taped to a window, you found the place where the dime just covered the moon and then measured the distance from you to the dime. The early Greeks did this for the moon and the sun and found what mathematical relationship?
8. Although there are no lakes or rivers on the moon, is there any indication of water on the moon? Explain.
9. Circle the things that have not been found on the moon: gravity, air pressure, lakes, wind, craters, mountains, sedimentary rocks, carbon.
10. Equipment left on the moon by astronauts allows scientists to measure the distance from the moon to the earth. Is the moon getting closer to the earth, getting farther away from the earth, or staying the same?

53

WHAT DID YOU LEARN?

1. Does the moon orbit the earth while the earth orbits the sun? *Yes*
2. How long does it take the earth to make one complete trip around the sun? About how long does it take the moon to make one complete trip around the earth? *The earth makes one complete trip around the sun in about 365 days. The moon makes one complete trip around the earth in about a month.*
3. Both the earth and the moon rotate on their axis. How long does it take for the earth to make one complete rotation? About how long does it take for the moon to make one complete rotation? *The earth makes one complete rotation on its axis in 24 hours. The moon makes one complete rotation on its axis in about 30 days.*
4. Why is only one side of the moon seen from the earth? *Because the moon only makes one rotation of its axis every month.*
5. Sometimes only part of the moon is visible in the night sky from the earth. What determines which part of the moon is visible at night? *From the earth, we can only see the part of the moon that is illuminated by the sun.*
6. How does the earth's gravitational pull affect the moon? How does the moon's gravitational pull affect the earth? *The earth's gravitational pull holds the moon in its orbital path around the earth. The moon's gravitational pull, combined with the sun's gravitational pull, causes the ocean tides.*
7. What does it mean to say the angular size of the moon and the angular size of the sun are nearly equal? *It means that both the moon and the sun appear to be the same size in the sky.*
8. Although there are no lakes or rivers on the moon, is there any indication of water on the moon? Explain. *An area at the moon's South Pole has been explored and found to have a large supply of water in the form of ice. This could be a source of drinking water and could also be a possible source of hydrogen for rocket fuel.*
9. Circle the things that have not been found on the moon: gravity, air pressure, lakes, wind, craters, mountains, sedimentary rocks, carbon. *Air pressure, lakes, wind, sedimentary rocks, and carbon*
10. Equipment left on the moon by astronauts allows scientists to measure the distance from the moon to the earth. Is the moon getting closer to the earth, getting farther away from the earth, or staying the same? *The moon is receding from the earth at the rate of about two inches per year.*

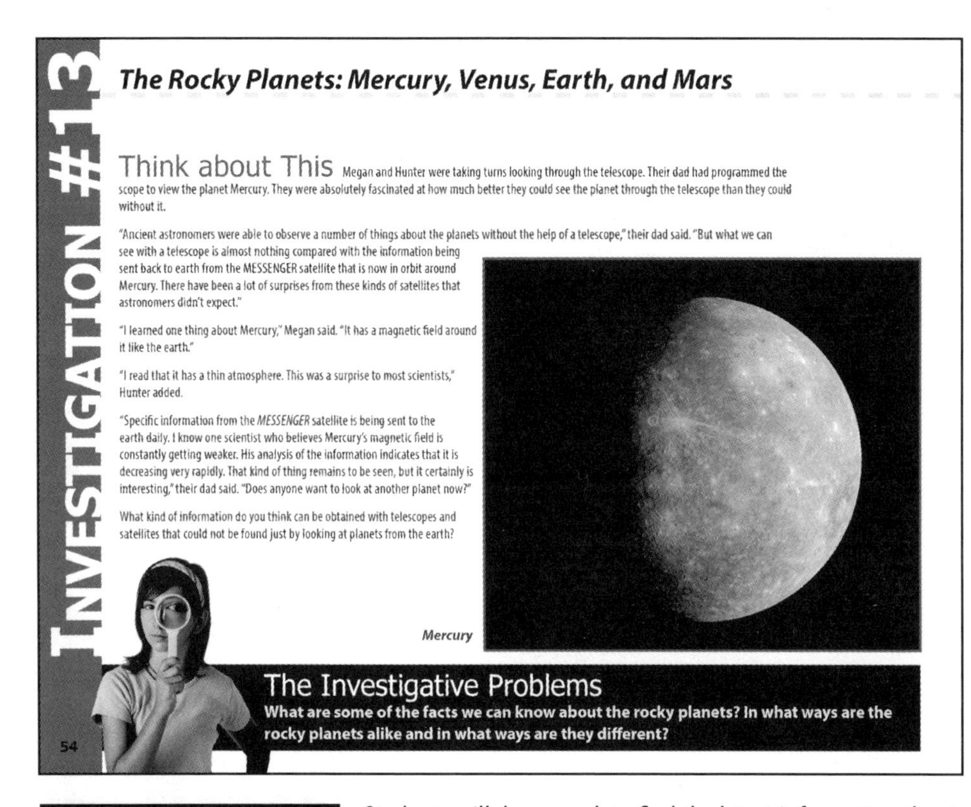

INVESTIGATION #13

The Rocky Planets: Mercury, Venus, Earth, and Mars

Think about This Megan and Hunter were taking turns looking through the telescope. Their dad had programmed the scope to view the planet Mercury. They were absolutely fascinated at how much better they could see the planet through the telescope than they could without it.

"Ancient astronomers were able to observe a number of things about the planets without the help of a telescope," their dad said. "But what we can see with a telescope is almost nothing compared with the information being sent back to earth from the MESSENGER satellite that is now in orbit around Mercury. There have been a lot of surprises from these kinds of satellites that astronomers didn't expect."

"I learned one thing about Mercury," Megan said. "It has a magnetic field around it like the earth."

"I read that it has a thin atmosphere. This was a surprise to most scientists," Hunter added.

"Specific information from the *MESSENGER* satellite is being sent to the earth daily. I know one scientist who believes Mercury's magnetic field is constantly getting weaker. His analysis of the information indicates that it is decreasing very rapidly. That kind of thing remains to be seen, but it certainly is interesting," their dad said. "Does anyone want to look at another planet now?"

What kind of information do you think can be obtained with telescopes and satellites that could not be found just by looking at planets from the earth?

Mercury

The Investigative Problems

What are some of the facts we can know about the rocky planets? In what ways are the rocky planets alike and in what ways are they different?

54

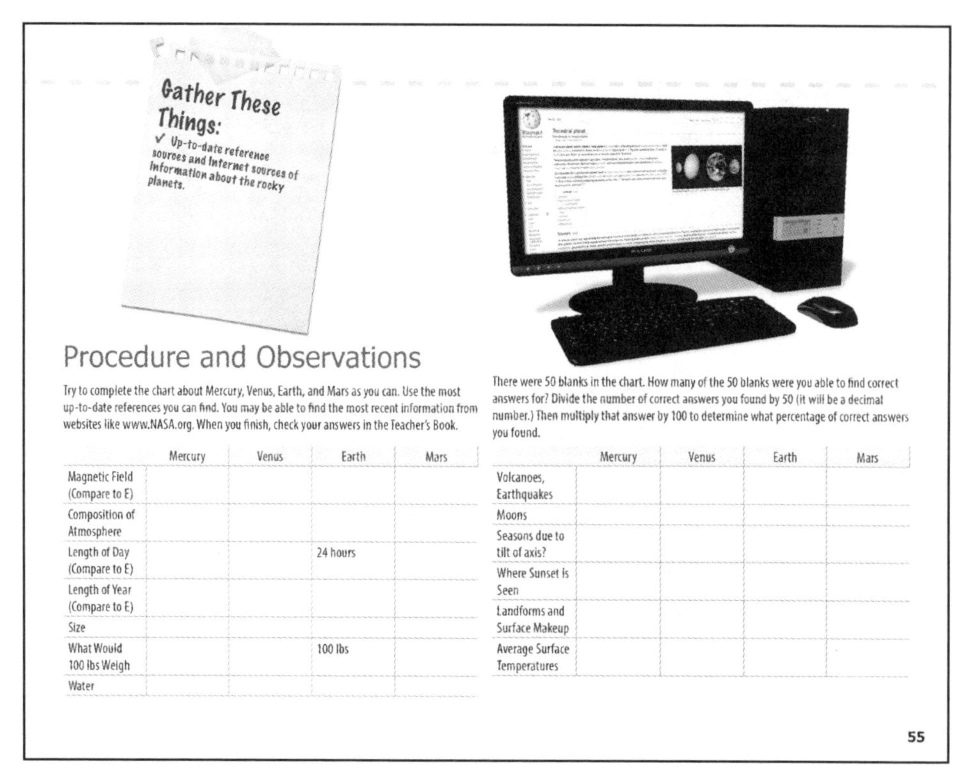

Gather These Things:

✓ Up-to-date reference sources and Internet sources of information about the rocky planets.

Procedure and Observations

Try to complete the chart about Mercury, Venus, Earth, and Mars as you can. Use the most up-to-date references you can find. You may be able to find the most recent information from websites like www.NASA.org. When you finish, check your answers in the Teacher's Book.

	Mercury	Venus	Earth	Mars
Magnetic Field (Compare to E)				
Composition of Atmosphere				
Length of Day (Compare to E)			24 hours	
Length of Year (Compare to E)				
Size				
What Would 100 lbs Weigh			100 lbs	
Water				

There were 50 blanks in the chart. How many of the 50 blanks were you able to find correct answers for? Divide the number of correct answers you found by 50 (it will be a decimal number.) Then multiply that answer by 100 to determine what percentage of correct answers you found.

	Mercury	Venus	Earth	Mars
Volcanoes, Earthquakes				
Moons				
Seasons due to tilt of axis?				
Where Sunset Is Seen				
Landforms and Surface Makeup				
Average Surface Temperatures				

55

OBJECTIVES

Students will do research to find the latest information about the rocky planets. They will compare magnetic field, atmosphere, length of day, length of year, size, gravitational pull, water, volcanoes, moons, seasons, where sunset is seen, landforms, temperature, and unusual features.

NOTE

Be sure students use up-to-date Internet sources or books. For example, older references will say that Mercury has no atmosphere, but newer data indicates that it does have a thin atmosphere. NASA webites can give you accurate up-to-date information from the space probes that have been sent up by NASA. Some NASA websites have been renamed and restructured, but up-to-date information is still available.

Be aware that NASA and many other organizations take a naturalistic, evolution-only approach to origins and consider anything that involves a supernatural aspect to be irrelevant. You might want to preview or combine chapter 20 in this book with this lesson.

Venus and Mars, as well as Jupiter, can be easily seen in the night sky most of the year with a telescope. Children find it very exciting to see celestial objects in the sky, especially if they know what they are seeing. There are several Internet sites that can help you find places to go to see celestial objects in the sky and identify these objects during each season. Be sure to take advantage of opportunities to do some sky watching.

Answers to Procedure and Observations chart on page 46.

The Science Stuff

Astronomers have been studying the first four rocky planets for hundreds of years. They have seen craters, mountains, plains, and valleys with their telescopes and studied chemicals with special instruments. They have even seen frozen polar caps on Mars that get bigger in a Martian winter and smaller in a Martian summer. But, many things can't be accurately determined from the earth's viewpoint. Since the space program began sending satellites into space, a great many new discoveries have been made, including some unexpected things.

The first phase of investigating the solar system began in 1959, when the Soviet Union launched a spacecraft that flew past the moon. During the next 30 years, astronauts landed on the moon and were able to explore it close up. Also, every planet in the solar system (except Pluto) was visited by at least one spacecraft.

Scientific exploration of the solar system by unmanned craft started up again in 1997 after a gap of nearly ten years. A spacecraft sent back images from the planet Jupiter and its moons; a robot explored the surface of Mars while another spacecraft went into orbit around the planet; and another spacecraft was sent toward Saturn. Some of the longstanding explanations and ideas about the solar system have been strengthened by these probes, but some ideas have been seriously called into question.

NASA's space shuttles were designed to be launched by rockets and then return to earth and land like airplanes. The first space shuttle was launched in 1981. An international space station has been built so that astronauts can be taken there and left for several months at a time to do space research. The U.S. space shuttle program has come to a close, but other countries will continue to provide transportation to and from the space station.

You were encouraged to use up-to-date references as you looked for information to complete the chart of the four rocky planets. New information is constantly being received from the satellites that are in space now.

The MESSENGER satellite went into orbit around Mercury in 2011 and began sending back information never known before. Scientists suspect the core of the planet is slowly cooling and becoming smaller, causing the whole globe to shrink. Many long and high cliffs on Mercury appear to be signs that the surface is crumbling as the planet buckles beneath it. One of the recent pictures of Mercury shows a large cliff running vertically for more than a hundred miles, along with similar patterns of other cliffs. Mercury has a thick metal core, making it unusually dense. It has a magnetic field. It only has a thin atmosphere, which could help explain why it is covered with craters.

Venus has a thick atmosphere of carbon dioxide and clouds of sulfuric acid. It is the hottest planet in our solar system, because its thick atmosphere traps the heat that comes from the sun. It is also one of the brightest lights in the night sky, except for the moon.

Unmanned satellites have actually landed on Mars and sent out land rovers to explore the surface and the atmosphere. The ground is very dry and contains a lot of iron rust. Mars has an atmosphere made of carbon dioxide gas in the summer, but in the winter the air next to the pole freezes and lies on the ground as a white solid. There is no sign of life, but there is frozen water deep underground in places.

Artist's rendering of MESSENGER orbiting Mercury

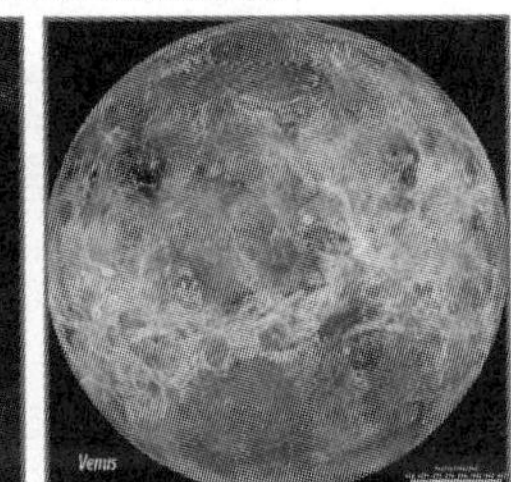

Venus

According to the nebula theory, the four planets closest to our sun should have formed under similar conditions, which mean they should have very similar properties except for their temperatures. They all orbit the sun in the same direction and on about the same plane. They are all rocky planets, with an abundance of iron. However, other conditions are very different. Study your chart to see all the ways they are alike and all the ways they are different.

One of the most difficult differences to explain in terms of the nebula theory is that Venus turns on its axis opposite to all the other planets. Mercury's heavy density is another feature that doesn't fit the nebula theory. Because Mercury spins in its axis very slowly, scientists were surprised to find that it has a magnet field. Most planets have lots of iron, but several other chemicals are different.

56

One hour timelapse photography

Voyager 2 program

Dig Deeper

Choose one of the space programs and do some research on it (MESSENGER satellite program, the Voyager 2 program, or another program you find interesting). Make a time-line of the program from the beginning until now. Include the process of planning and launching. Tell about the things that were observed and detected by the satellite.

There are belts of asteroids between Mars and Jupiter made up of large pieces of rocks and boulders. According to one hypothesis, these are the remains of a planet that exploded or disintegrated. There is also another ring of large rocky objects beyond Neptune, known as Trans-Neptunian Objects or TNOs. Pluto and Pluto's moons are a part of this area. See if you can find additional information about the asteroid belts and the TNOs. Write about this in an interesting way.

Making Connections

The constellations appear every year in precise predictable order, but early astronomers had difficulty finding a predictable pattern for the planets, which were also called stars. The Greek word for planet means "wanderer." From the earth's viewpoint, the planets sometimes seemed to move backward or move around in a circle. They don't really do this, but people on earth are looking at them from a moving platform, which make them appear to be moving in random ways. You can experience something similar to this if you are in a slowly moving vehicle and look out the window at another large vehicle where nothing else is visible. The other vehicle may appear to be moving backward.

What Did You Learn?

1. Which of the rocky planets have the strongest magnetic fields?
2. Why is Venus the hottest planet in our solar system?
3. Which of the rocky planets has liquid water on its surface?
4. On which of the rocky planets is there a sunset in the east?
5. Which of the rocky planets has the longest day? Which one has the shortest day?
6. Which of the planets does not have spring, summer, fall, and winter seasons?
7. In what year did the Soviet Union first launch a spaceship that flew past the moon?
8. Do all the rocky planets have an atmosphere? Are they all made up of the same gases?
9. What did the MESSENGER satellite discover about the planet Mercury that may indicate it is cooling and shrinking?

57

What Did You Learn?

1. Which of the rocky planets have the strongest magnetic fields? *Mercury and Earth*
2. Why is Venus the hottest planet in our solar system? *It is close to the sun and has a thick atmosphere that traps heat from the sun.*
3. Which of the rocky planets has liquid water on its surface? *Earth*
4. On which of the rocky planets is the sunset in the east? *Venus*
5. Which of the rocky planets has the longest day? Which one has the shortest day? *Venus has the longest day. Earth has the shortest day.*
6. Which of the rocky planets does not have spring, summer, fall, and winter seasons? *Mercury and Venus.*
7. In what year did the Soviet Union first launch a space ship that flew past the moon? *1959*
8. Do all the rocky planets have an atmosphere? Are they all made up of the same gases? *All the rocky planets have an atmosphere, but they differ in thickness, and they do not all have the same gases.*
9. What did the *MESSENGER* satellite discover about the planet Mercury that may indicate it is cooling and shrinking? *Many long and high cliffs on Mercury may be a sign that the planet is cooling and shrinking.*

INVESTIGATION #14

Planets: Mars and Martians

Think about This

The search for life on other planets or on moons is a major emphasis of the space program. Forms of life have been found to survive on earth under extreme conditions of temperature and pressure. This caused some scientists to believe there may be life on other planets or moons where the temperatures are extremely hot or cold or where air pressures are very different from that on earth. However, the thing that scientists look for as essential for life more than anything else is water. From the very beginning of the space program, there has been a great search for water.

A number of overly enthusiastic reporters have written headlines suggesting that life has been found on other planets in our solar system. It turns out that evidence for water (or some kind of organic material) was discovered in some places outside of the earth.

A few years ago, planets were discovered in other solar systems in our galaxy. Some of these planets were at a distance from their sun (not too close) where the temperatures were below the boiling point of water. Some of these planets were at a distance (not too far away from their sun) where the temperatures were not always below the freezing point of water. Scientists reasoned that under these conditions, water might be present in a liquid form. And, if water was present, then living things might also be present. Water in space is not that unusual. Concluding that water means life, of course, is really bad logic. But the great search for water in space goes on.

Does the presence of water on another planet or moon mean that there will also be living things there? What do you think about this kind of reasoning?

Many astronomers believed that there was life on mars because water exists presently in the form of ice. This planet might have had abundant water in the past, but with much research including robotic rovers sifting soil samples there is no evidence of any life existing in the past.

The Investigative Problems

Does water exist in space in places other than the earth? Does the presence of water on other planets or moons mean that there are forms of life there?

Procedure and Observations

For this lesson, you will need to do some Internet research or find some other reference materials. Try to find at least ten headlines from newspapers or magazines having to do with "extraterrestrial life" or life in space or with finding conditions that would support life. Try to find both old and recent articles. Copy the headlines and any sub-headlines. Give the name of the newspaper or magazine and the date it was published. Write your thoughts about whether the headlines are justified or not.

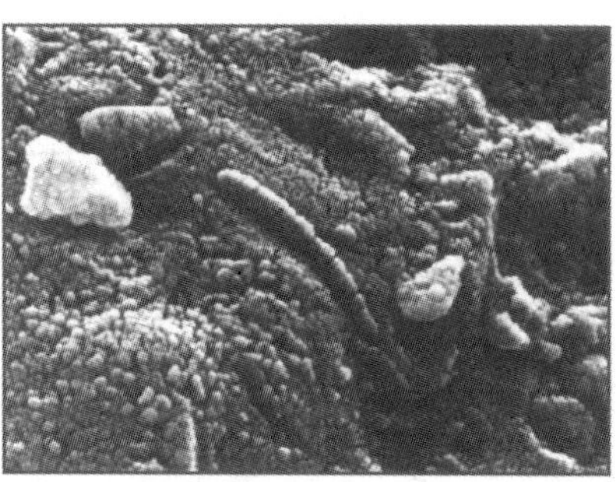

At one time scientists thought that this magnified image from a Martian meteorite showed fossilized bacteria-like organisms. Recent space probes to Mars no longer support that idea.

Objectives

1. The search for life on other planets or on moons is a major emphasis of the space program.
2. The thing that scientists look for as essential for life more than anything else is water. From the very beginning of the space program, there has been a great search for water.
3. A few years ago, planets were discovered in other solar systems in our galaxy. Some of these planets were the right size and distance from their sun to possibly have liquid water. Scientists reasoned that if water was present, then living things might also be present. However, none of the space probes have found any evidence of living things.
4. Scientists who only accept natural explanations for everything that happens in the present and in the past are known as naturalists. All scientists look for natural explanations as much as possible, but creation scientists recognize that when it comes to how matter, stars, planets, life, and mankind came to be, random natural processes cannot provide adequate explanations.

Note

An Internet search of "extraterrestrial life" will yield several thousand results. Many of these will be highly exaggerated or fictitious accounts. Several popular science magazines have also published articles about extraterrestrial life. You can find several articles referenced from more professional publications, such as http://discovermagazine.com/topics/space/extraterrestrial-life. However, many of these articles will probably be biased toward focusing on evidence that might indicate some kind of extraterrestrial life exists. Students should try to find some tabloid-type headlines and some more professional headlines. They should look for both recent and older headlines.

Try to help students realize that in spite of all of these articles, no one has discovered living things anywhere except on earth. Most of the articles students will find have to do with the discoveries of finding water, possible places where water might exist, or chemicals containing carbon, hydrogen, oxygen, nitrogen, or other elements found in living things. Help students to recognize that SEITA researchers are basing their hypotheses on a belief that the evolution of all living things must be true. Help students understand that these articles and hypotheses are based on an assumption that life on earth evolved from natural process and random events. Evolutionists reasoned that if life evolved on earth where all of these conditions exist, then in our huge universe, there must be other places where life evolved. Also, help students realize that when people are looking for a particular outcome, they are already prejudiced toward finding it. Most scientific researchers use methods to try to neutralize their prejudices, but for areas like

Mars

The Science Stuff

Mars is just close enough to the earth to see some details through telescopes. Of all the planets, Mars has been the one where people have imagined there might be alien living things. Mars is close enough to see different features through a good telescope on Earth. It has always been fascinating to see the changing seasonal features and wonder what in the world is on that planet. Mars has white areas around its north pole and its south pole that get larger during the winter season and smaller during the summer season. This is primarily frozen carbon dioxide in the winter, but there seems to be frozen water deeper underground.

During the mid 1800s, a well-known scientist proposed that Mars had seas, land, and possibly forms of life. As telescopes improved, eager observers speculated about all kinds of things they thought were happening on Mars. Much was written about so-called Martian canals, thought to be the work of alien beings. Some popular science fiction books like Jules Verne's *Around the Moon* (1870) and H.B. Wells' *The War of the Worlds* (1898) helped to make the idea of space aliens more believable.

In 1938, there was panic in some places because a radio drama version of *The War of the Worlds* appeared to be the "real thing." Conspiracy theories, alien abductions, UFOs, and flying saucers also popped up. It quickly became clear after the space program began that neither the moon nor other bodies of the solar system are home to intelligent life, although some evolutionary scientists continue to look for bacteria and such in the solar system.

The idea of ancient astronauts who have visited the earth in the remote past began to appear in science fiction books in early 20th century. But, more recently, evidence for a very old advanced civilization has been given serious consideration by respected scientists. The authors of this book do not believe in space aliens, but they believe there may have been early advanced civilizations on earth. The evidence for very old technologies should be considered and studied, because the

60

pre-Flood people were very intelligent and probably had many technologies. Some of this information would have been passed on to Noah and his family who survived the Flood.

The space program's search for life on other planets is a part of the search for the origin and evolution of everything. Many scientists insist that there must be a natural explanation for how everything began. In the process of looking only for natural explanations, they also reject the possibility that everything began with a supernatural creation.

At one time, most Americans and Europeans, including famous scientists, believed that God supernaturally created the heavens and the earth. This was accepted as one of the things that God chose to reveal to us in His Word, rather than something people figured out by using reason and logic.

Then, sometime in the late 1500s, things began to change — slowly at first, but becoming more and more widespread. People began to think that the best way to find out what is true was not to look at what God revealed in His Word, but to use human reason. This shift from "revelation to reason" is known in history as the Enlightenment Period or modernism.

Sometime during the 1800s, more and more scientists began to reject the idea that God supernaturally created the heavens and the earth. They began assuming that there must be natural explanations for how the universe got started. Naturalism is a philosophy, but it has a big effect on science, especially historical sciences.

Darwin's theory for the evolution of living things in the mid-1800s fit well into these ideas. The big-bang and the nebula theories also provided natural explanations for how everything came to be.

Making Connections

It might surprise you to know that scientific polls for the last several years have shown than the majority of Americans still believe that God designed and created the earth and mankind. Our culture has certainly been influenced by the Enlightenment ideas, but they have by no means replaced the Bible or Christian faith.

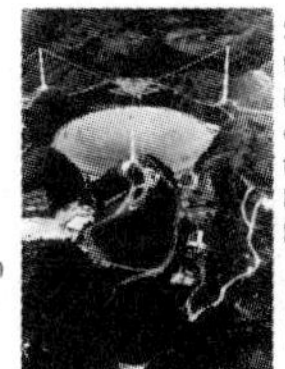

Some astronomers are convinced that there are intelligent beings in outer space and have conducted serious research to find alien life forms in a program known as SETI (Search for Extraterrestrial intelligence).

Image of the Arecibo Observatory in Puerto Rico with world's largest dish that is sometime used for SETI research.

Visual summary of White House position on extraterrestrials: no evidence yet, search being conducted by

1. looking for extrasolar planets (Kepler mission)

2. listening for signals (SETI, illustrated by Allen array)

3. robotic exploration of the Solar System (illustrated by Curiosity mars rover) NASA/JPL-Caltech

Searching for ET, But No Evidence Yet

Dig Deeper

Astronomers often propose evolutionary theories about things in the universe based on things they cannot actually observe. Try to find evidence for Oort clouds or the birth of a star. Did you find evidence for them or just an explanation for what might have happened?

What is SETI and what kind of research do they do?

Look at the chart you completed in the previous lesson. List at least five facts that agree with the idea of a young solar system better than it does with one that proposes that the solar system is billions of years old. Explain if you need to.

What Did You Learn?

1. Before the space program began, why did some scientists believe there were living things on Mars?
2. Have space satellites and probes found evidence of life on any of the planets?
3. Some people believe the earth was visited by aliens from outer space who taught ancient civilizations their advanced technologies. What is another explanation for this?
4. Is one of the goals of the space program to find more about the origin and evolution of everything?
5. Do many scientists reject a supernatural explanation for how everything began because they believe there must be a natural explanation for this?
6. Briefly tell what happened during the Enlightenment Period of history.
7. Give one thing proposed by naturalism.
8. What kind of research does SETI do?

61

evolution and creation, this may be impossible. It would be an error to overlook the possibility that the universe was planned rather than happening by chance.

What Did You Learn?

1. Before the space program began, why did some scientists believe there were living things on Mars? *Mars is close enough to observe changes that occur in the winter and other seasons with telescopes. During the mid 1800s, a well-known scientist proposed that Mars had seas, land, and possibly forms of life. There were lots of speculations about what might be happening on Mars.*
2. Have space satellites and probes found evidence of life on any of the planets? *No*
3. Some people believe the earth was visited by aliens from outer space who taught ancient civilizations their advanced technologies. What is another explanation for this? *Some scientists have been considering evidence that there were once advanced civilizations. The pre-Flood people were probably very intelligent and had developed many technologies. Some of this information would have been passed on to Noah and his family who survived the Flood.*
4. Is one of the goals of the space program to find more about the origin and evolution of everything? *Yes*
5. Do many scientists reject a supernatural explanation for how everything began because they believe there must be a natural explanation for this? *Yes*
6. Briefly tell what happened during the Enlightenment Period of history. *People began to think that the best way to find out what is true was not to look at what God revealed in His Word, but to use human reason. There was a shift from "revelation to reason."*
7. Give one thing proposed by naturalism. *Naturalism rejects all ideas that involve supernatural events, such that God supernaturally created the heavens and the earth. They assume that there must be natural explanations for how the universe got started.*
8. What kind of research does SETI do? *SETI stands for Search for Extra-Terrestrial Intelligence. Their research involves looking for intelligent life in space.*

INVESTIGATION #15

The Jovian Planets: Jupiter, Saturn, Uranus, and Neptune

Think about This

Jonah and Sarah studied a map of the United States with their parents. They were planning to drive a camper from Miami, Florida, to Anchorage, Alaska, during the upcoming summer. There were other places along the way they also wanted to see. His dad said the trip was about 5,000 miles and their goal was to average 300 miles a day. Sarah did some quick calculations and was surprised to find that the trip would take almost 17 days for a one-way journey.

"It sure doesn't look that far on this map," Jonah commented.

"No, it doesn't," his dad replied. "If you think that's a long trip, how would you like to travel through space from the earth to Jupiter? It took one spacecraft six years to get there.

"NASA technicians launched an unmanned spacecraft named Galileo from the earth in October of 1989. It finally arrived at Jupiter in December of 1995 after exploring asteroids on the way. For the next several years, it orbited Jupiter and explored both the planet and its moons until its fuel began to run low. They thought about what to do with the last supply of fuel.

"Scientists had long been wondering what was beneath the red, brown, orange, and yellow layers of gases. They decided to crash Galileo into Jupiter and let it send back signals as long as it could. They thought they could determine if it was made completely of gases or if there was a solid core in the middle of it."

Artist's impression of Galileo flying past Io, one of the moons of Jupiter.

"So, what did they find out from the crash test?" Sarah asked.

"On its final mission, Galileo plunged into the atmosphere of Jupiter. It was able to travel thousand of miles into the atmosphere before it finally stopped sending signals back to Earth. As a result of this mission, astronomers learned that Jupiter's atmosphere is at least several thousand miles thick, but it may be deeper than that."

How long do you think it would take to reach the planet Neptune? Do you think astronauts will ever be able to travel to any of the giant planets? Do you think they will be able to land on one of them?

The Investigative Problems

What are some of the properties of the giant planets and their moons?

62

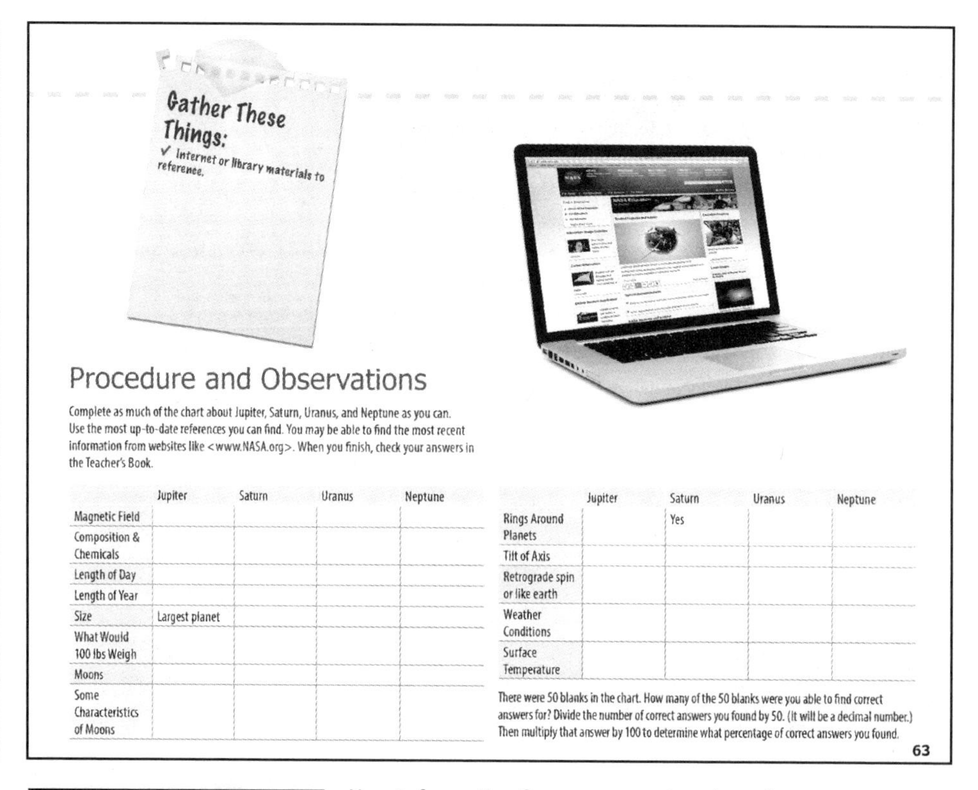

Gather These Things:

✓ Internet or library materials to reference.

Procedure and Observations

Complete as much of the chart about Jupiter, Saturn, Uranus, and Neptune as you can. Use the most up-to-date references you can find. You may be able to find the most recent information from websites like <www.NASA.org>. When you finish, check your answers in the Teacher's Book.

	Jupiter	Saturn	Uranus	Neptune
Magnetic Field				
Composition & Chemicals				
Length of Day				
Length of Year				
Size	Largest planet			
What Would 100 lbs Weigh				
Moons				
Some Characteristics of Moons				

	Jupiter	Saturn	Uranus	Neptune
Rings Around Planets		Yes		
Tilt of Axis				
Retrograde spin or like earth				
Weather Conditions				
Surface Temperature				

There were 50 blanks in the chart. How many of the 50 blanks were you able to find correct answers for? Divide the number of correct answers you found by 50. (It will be a decimal number.) Then multiply that answer by 100 to determine what percentage of correct answers you found.

63

OBJECTIVES

Students will do research to find the latest information about the giant gas planets. They will compare magnetic field, atmosphere, length of day, length of year, size, gravitational pull, weather conditions, moons, rings, kind of rotation, temperature, and unusual features.

NOTE

New information from space exploration often outpaces information in textbooks. Encourage students to research the most up-to-date resources they can find. They should not be surprised if more recent information is different from what they find in older books.

Information coming from satellites in space often changes. Students should try to find the most up-to-date-information they can. NASA has several good sources of information.

*These answers will vary considerably depending on which moons students choose to mention.

**The answers may also vary according to whether the temperatures refer to tops of clouds, lower clouds, average temperatures, effective temperatures, or some other measure. Look for reasonable answers. Metric or English numbers may be used.

Answers to Procedure and Observations chart on page 47.

The Science Stuff

Jupiter, Saturn, Uranus, and Neptune, known as the Jovian planets, are the solar system's giant gas planets. It is uncertain whether or not these planets have a rocky core. We know that two of their main elements are hydrogen and helium. They also contain other compounds such as methane and ammonia. The gases that make up the planets are very active, producing mega storms and winds. There is a huge cyclone storm on Jupiter, known as the Great Red Spot, and another cyclone storm on Neptune, known as the Great Dark Spot.

A montage of Jupiter and its four largest moons

The gases also give Jupiter and Saturn their beautiful layers of colors, mostly reds, browns, oranges, and yellows. Uranus and Neptune are shades of deep blue or blue-green.

All of the gas planets have multiple rings and moons around them. Saturn's rings are the largest and can be observed from the earth with a good telescope. Some of the moons can be seen from the earth with telescopes, but many of the rings and moons were discovered by unmanned spacecraft that passed nearby.

The gas giant planets are much larger than the rocky planets. They also have stronger gravitational attractions. Jupiter is about one thousand times larger than the earth.

Uranus is unusual in that it is tilted on its side, and its large moons orbit around its equator. It also rotates backward, like Venus.

The moons and the rings of the four outer giant planets are almost as fascinating as the planets themselves. They are certainly the hardest to explain in terms of the nebula theory and billions of years.

All the moons around the gas planets are rocky objects, but some of them have ice, water, other liquids, or frozen gases. All of the inner moons of Uranus are thought to be about half water or ice and half rock. Astronomers believe there is a large ocean of water or slushy ice beneath a surface layer of ice on Jupiter's moon, Europa. Some of Jupiter's other moons also appear to contain frozen water.

Triton (Neptune's largest moon) has ice volcanoes that shoot material 5 miles (8 km) above the moon's surface. This is probably a mixture of liquid nitrogen, methane, and dust that falls back to the surface as a frozen snow-like substance. Enceladus (with Saturn), also seems to have active ice volcanism caused by heat within the moon. One of the space probes observed warm fractures on Enceladus where melting ice was evaporating and forming a huge cloud of water vapor. Although Triton is one of the coldest objects in the solar system -240 degrees Celsius, there apparently is enough internal heat to expel gases from the ice volcanoes.

The moon Io (with Jupiter) has the most active volcanoes in the solar system. Its volcanic eruptions are caused by hot magma below the surface.

Some of the moons turn on their axis in a clockwise motion and some of them turn in a counter-clockwise motion. Triton circles Neptune in a direction opposite to Neptune's rotation. This motion acts as a drag on Triton, causing it to get closer and closer to Neptune. It is predicted to eventually break apart from the gravitational forces on it and form another ring around Neptune. Some of Saturn's moons, including Phoebe, orbit the planet in a direction opposite that of Saturn's larger moons.

Most of the moons do not have an atmosphere, but there are a few that do. Saturn's largest moon, Titan, has an atmosphere that is about 95 percent nitrogen with small amounts of methane, even though it is smaller than the earth.

Many of these observations are hard to explain in terms of billions of years. Outer space is very cold and after a few billion years, we would expect most celestial bodies without atmospheres to become frozen solid.

The moons and planets with irregular spins and orbits cannot be explained by the nebula hypothesis. Astronomers try to explain them by gravitational captures or collisions with other objects in space, but there are additional problems with these explanations.

64

Dig Deeper

Choose one of the moons that orbit a gas giant planet and find all the information you can about it. Display this information and include pictures or diagrams.

Choose one of the gas giant planets and find information about the planet that was learned in the past ten years. Display this information and include pictures or diagrams.

Making Connections

Several years ago, astronomers began to discover other solar systems in the universe where planets orbited a star. They mainly use indirect methods to detect these faraway solar systems and their planets, although recently some large gas giant planets have actually been observed. Scientists assumed the planets would be similar to those in our solar system with rocky planets closest to their star and large gas planets farther away from their star. Surprisingly, many of large gas planets orbit very close to their star.

What Did You Learn?

1. Which of the Jovian planets have a strong magnetic field?
2. Which is the largest Jovian planet?
3. Which Jovian planet has the most moons?
4. Which Jovian planet has the longest day? Which one has the shortest day?
5. Which Jovian planets have rings around them?
6. Which Jovian planet is most tilted on its axis?
7. Which Jovian planet spins on its axis opposite to the way the earth spins?
8. Jovian planets are made up of what two main elements?
9. Where are there huge cyclone storms on Jovian planets?
10. What is one thing that is unusual about Jupiter's moon Europa?
11. What is one thing that is unusual about Neptune's moon Triton?
12. Name two other moons that orbit Jovian planets and give some facts about them.

These colors indicate the varying heights and compositions of cloud layers on Saturn. You can also see the moons of Saturn, Tethys (upper right), and Dione (lower left).

65

WHAT DID YOU LEARN?

1. Which of the Jovian planets have a strong magnetic field? *Jupiter and Saturn have the strongest magnetic fields. (Some astronomers would consider Uranus and Neptune as having strong magnetic fields.)*
2. Which is the largest Jovian planet? *Jupiter*
3. Which Jovian planet has the most moons? *Jupiter (Some astronomers might consider Saturn as having the most because some of the moon-like objects are listed as provisional.)*
4. Which Jovian planet has the longest day? Which one has the shortest day? *The Jovian planet with the longest day is Uranus. The one with the shortest day is Jupiter.*
5. Which Jovian planets have rings around them? *Saturn has the most rings around it, but Jupiter, Uranus, and Neptune also have rings around them.*
6. Which Jovian planet is most tilted on its axis? *Uranus is tilted the most on its axis.*
7. Which Jovian planet spins on its axis opposite to the way the sun and the earth spins? *Uranus*
8. Jovian planets are made up of what two main elements? *Hydrogen and helium*
9. Where are there huge cyclone storms on Jovian planets? *Jupiter's Great Red Spot and Neptune's Great Dark Spot are thought to be cyclone storms.*
10. What is one thing that is unusual about Jupiter's moon Europa? *Its surface is covered with ice and it's thought to consist of large amounts of slushy water beneath.*
11. What is one thing that is unusual about Neptune's moon Tritan? *It has ice volcanoes and reaches temperatures of -400° F.*
12. Name two other moons that orbit Jovian planets and give some facts about them. *This is student choice.*

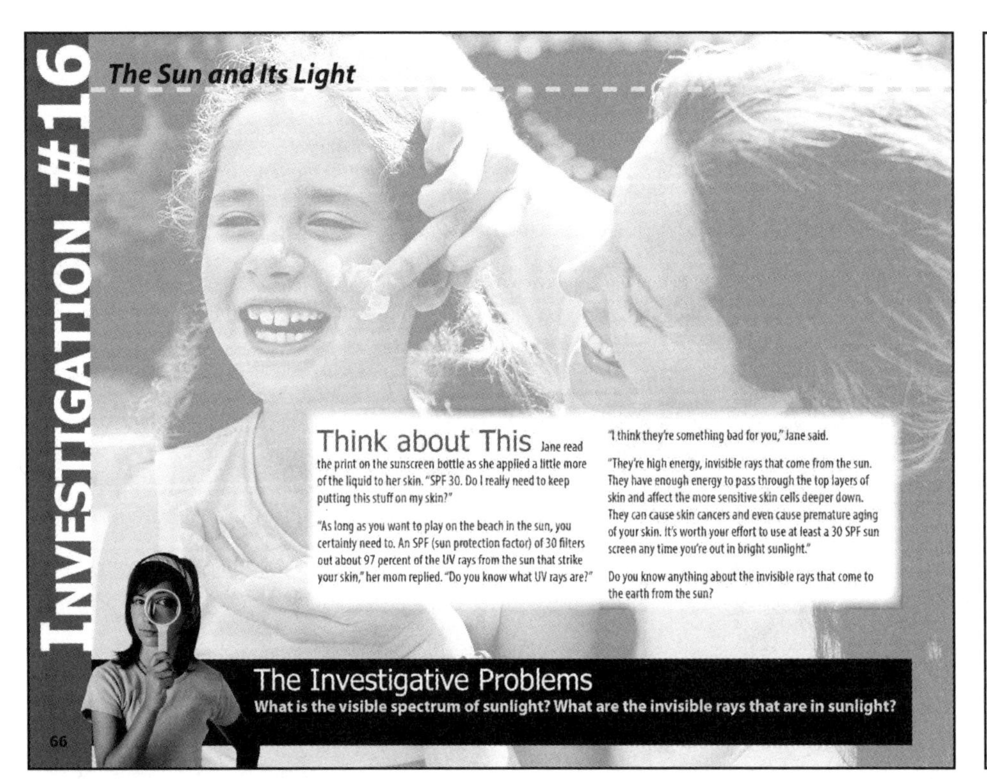

INVESTIGATION #16

The Sun and Its Light

Think about This

Jane read the print on the sunscreen bottle as she applied a little more of the liquid to her skin. "SPF 30. Do I really need to keep putting this stuff on my skin?"

"As long as you want to play on the beach in the sun, you certainly need to. An SPF (sun protection factor) of 30 filters out about 97 percent of the UV rays from the sun that strike your skin," her mom replied. "Do you know what UV rays are?"

"I think they're something bad for you," Jane said.

"They're high energy, invisible rays that come from the sun. They have enough energy to pass through the top layers of skin and affect the more sensitive skin cells deeper down. They can cause skin cancers and even cause premature aging of your skin. It's worth your effort to use at least a 30 SPF sun screen any time you're out in bright sunlight."

Do you know anything about the invisible rays that come to the earth from the sun?

The Investigative Problems

What is the visible spectrum of sunlight? What are the invisible rays that are in sunlight?

66

Procedure and Observations

Part A.

Read the label on different bottles of sunscreen. Make a chart comparing the different bottles. Include brand name, SPF rating, the ingredients, and any other features mentioned.

Part B.

Use the sun, a projector, or a strong flashlight as a source of light. Shine the light on something white. Place the prism in the path of the beam of light and turn it slightly back and forth. Carefully observe the colors that form on the white surface. What do you observe? Make a sketch on your paper of the bands of color you saw. Name each color and show the order in which they appear on the paper. If you have trouble seeing the spectrum clearly, hold the prism next to your eye and look at a light bulb (that is on), and name the colors you see from the light. (Do NOT look at the sun!)

The Science Stuff

The SPF rating on a bottle of sunscreen lotion gives you the sun protection factor you can expect if your skin is covered with this lotion. There's not much difference in the effectiveness of a SPF of 30 and an SPF of 100. They both filter out most of the harmful UV rays as long as the lotion remains on a person's skin.

The band of colors that you observed is called a color spectrum. A prism bends light as it passes through, and divides the light into a spectrum of red, orange, yellow, green, blue, indigo, and violet. Some wavelengths are bent more than others. The longest red waves are bent the least. The shorter violet rays are bent the most. The band of colors produced by a prism are the same ones you see in a rainbow.

There are also invisible wavelengths that come from the sun. Invisible waves that are a little longer than red are called infrared waves. They can be felt as heat waves. Invisible waves that are a little shorter than violet are called ultraviolet waves or UV waves. Radio waves are very long waves, and x-rays and gamma rays are very short waves. An important point to remember is that the shorter wavelengths have more energy than the longer ones. They are also the most dangerous waves. The earth is protected from many of these dangerous rays by the earth's magnetic field.

67

OBJECTIVES

1. When light rays pass through a prism, the visible rays separate into a spectrum of colors from red, the longest, to violet, the shortest.
2. The invisible rays can be detected with other equipment and are often photographed with special photography.
3. The entire range of waves — visible and invisible — that come from the sun are known as electromagnetic waves. All electromagnetic waves travel at the same speed and can travel through space, but they can differ in their wavelengths and in how much energy they have.
4. Energy is released by nuclear fusion as hydrogen nuclei are converted into helium nuclei. This energy emerges as light, heat, and other cosmic radiations.

NOTE

The sun is an amazing object. Scientists from very early civilizations recognized its properties and its effects on earth. They were able to measure the seasons and make clocks and calendars based on the patterns of its movement. At first their observations were scientific, but later on, their fascination with astrology turned to worship of the sun and the stars in spite of stern warnings from God not to do this. An interesting study is to compare civilizations that worshiped the sun.

WHAT DID YOU LEARN?

1. Electromagnetic waves travel through space at what speed? *They travel through space at the speed of light.*
2. What are three ways in which electromagnetic waves can differ from each other? *They differ in wavelengths, frequencies, and amount of energy.*
3. Name three kinds of electromagnetic waves that are longer than visible light. *Electromagnetic waves such as infrared, radar, TV, FM radio, and AM radio are all longer than visible light.*

Electromagnetic waves also vibrate as they move. The rate at which each kind of wave vibrates is known as its frequency. Short waves have a high frequency and long waves have a low frequency.

Both visible and invisible waves come from the sun and from other stars. The entire range of waves that come from the sun are known as electromagnetic waves and are listed on the chart below. All electromagnetic waves travel at the speed of light and can travel through space. They differ in their wavelengths, frequencies, and the amount of energy they have.

Notice that sound waves are not electromagnetic waves and are not included in the chart. They cannot travel through space. They also differ in the way in which they vibrate.

Electromagnetic Spectrum

Increasing energy

Increasing wavelength

0.0001nm 0.01nm 10nm 1000nm 0.01cm 1cm 1m 100m

Gamma rays | X-rays | Ultra-violet | Infared | Radio waves

Radar TV FM AM

Visible light

400nm 500nm 600nm 700nm

Astronomers use an instrument called a spectrograph to produce a spectrum of visible light from the sun. This is much like what you did by shining a light through a prism. They can also produce spectra of light from stars that are trillions of miles away and identify the elements that are on these stars.

Incandescent objects, like stars, shine because the elements in them are glowing with heat. Whenever an element is heated until it becomes incandescent, it will produce a spectrum that is different from any other element. Light coming from elements on a distant star passes through cool gases, where energy is absorbed. This absorption of energy shows up on a spectrum as dark lines. These dark-line spectra can be used to identify each element present. Astronomers who are trained to analyze spectral lines cannot only identify elements that are present, but they can determine information about the temperature and density of the elements and the magnetic field of the star.

Each element present in a star will produce lines in different places in the dark-line spectrum.

Here are dark-line spectra for two elements:

Calcium

Hydrogen

Astronomers study galaxies and other celestial bodies by examining them with all kinds of equipment used to detect electromagnetic waves: visible light, x-rays, ultraviolet, infrared, and radio waves. Each type of radiation reveals different features.

Making Connections

A bright-line spectrum shows the same pattern of lines found on a dark-line spectrum, but the incandescent elements are not cooled in the process. Instead of dark lines, the lines are bright.

The most dangerous electromagnetic waves are the ones with very short wavelengths, because they have the greatest amounts of energy. You may know that x-rays have enough energy to pass through soft tissues in your body, but are blocked by hard substances like bones and teeth. People who work with x-rays on a daily basis take precautions to avoid too much exposure to these rays.

Infrared light and ultraviolet light can be photographed using special equipment even though you can't ordinarily see them. They can even be used to make photographs at night and in cloudy conditions. Police and military personnel often use night-vision infrared goggles to see heated objects at night. The heat from a person being watched gives an image of the person that couldn't ordinarily be seen at night.

Ultraviolet light is given off by a "black light," which can be bought from many places. Ultraviolet light causes florescent objects to glow in the dark. This equipment has been used to make photographs in space to show conditions that would not show up with cameras based on visible light.

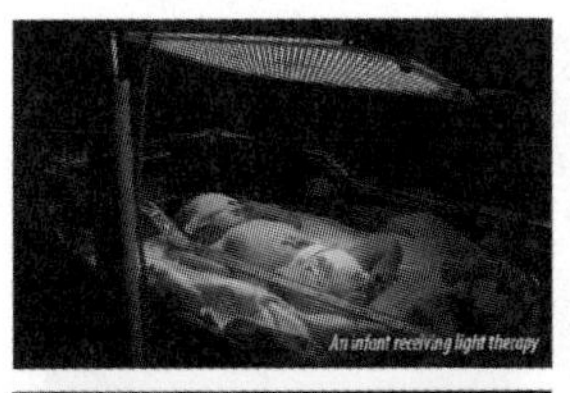

An infant receiving light therapy

Infrared light photograph

Aviator's Night Vision Imaging System

What Did You Learn?

1. Electromagnetic waves travel through space at what speed?
2. What are three ways in which electromagnetic waves can differ from each other?
3. Name three kinds of electromagnetic waves that are longer than visible light.
4. Name three kinds of electromagnetic waves that are shorter than visible light.
5. Which are more dangerous — very short electromagnetic waves or very long electromagnetic waves?
6. Do very short electromagnetic waves have high frequencies (vibrate faster) or low frequencies (vibrate slower)?
7. What kind of electromagnetic waves are filtered out by sunscreen lotion?
8. How do sound waves differ from light waves and other kinds of electromagnetic waves?
9. What kind of wave is detected by night-vision goggles?
10. What instrument do astronomers use to produce a spectrum of visible light from the sun and other stars? What information about these stars can they obtain from studying these spectra?

Dig Deeper

If you have access to photography equipment that allows you to photograph infrared and ultraviolet light, make some pictures. Then make pictures of the same objects using an ordinary camera and compare the pictures. Alternatively, you might be able to look through an infrared scope at night and compare what you see at night with what you see in daylight.

Use different colored filters to grow plants to determine under which color the plants grow best. Be sure to keep everything you do the same except for the light. Grow one set of plants in ordinary light and another set of plants in a light that is covered up with a colored filter. You could also try growing plants under an ultraviolet light or an infrared light.

4. Name three kinds of electromagnetic waves that are shorter than visible light. *Electromagnetic waves such as gamma rays, x-rays, and ultraviolet waves are all shorter than visible light.*

5. Which are more dangerous — very short electromagnetic waves or very long electromagnetic waves? *Very short electromagnetic waves are more dangerous than the long waves.*

6. Do very short electromagnetic waves have high frequencies (vibrate faster) or low frequencies (vibrate slower)? *Very short electromagnetic waves have high frequencies and vibrate faster than the long waves.*

7. What kind of electromagnetic waves are filtered out by sunscreen lotion? *Sunscreen lotion is designed to filter out ultraviolet waves (UV waves).*

8. How do sound waves differ from light waves and other kinds of electromagnetic waves? *Sound waves cannot travel through space. Electromagnetic waves can travel through space. Sound waves travel at speeds of about 1,100 feet per second. Electromagnetic waves travel at speeds of about 186,000 feet per second. They also differ in how they vibrate.*

9. What kind of wave is detected by night-vision goggles? *Infrared waves (heat waves).*

10. What instrument do astronomers use to produce a spectrum of visible light from the sun and other stars? What information about these stars can they obtain from studying these spectra? *Spectroscopes produce spectra of visible light from the sun and other stars that enable scientists to identify the elements that are present. Other information may also be obtained from scientists who know how to interpret the spectra.*

INVESTIGATION #17

The Sun and the Earth Relationship

Think about This

Throughout history, many groups of people have worshiped the sun, moon, and stars. Sometimes the sun god was considered the chief god and sometimes other beings were believed to be more powerful than the sun god. The stories about these gods are bizarre. The gods were thought to be jealous and vengeful and only gave blessings to people of whom they approved and had earned special favors. Ra (Eqyptian), Helios (Greek), and Apollo (Roman) were a few of the names given to sun gods by different cultures. There were many other names for sun gods around the world.

The Jewish people of the Old Testament stood out in great contrast to these other cultures, because they only believed in the one true God, who loved and protected His people. The laws given to them included strong warnings not to worship the sun, moon, and stars or any other kind of false gods

When we hear of people in different parts of the world who still believe in idol worship, we tend to think they are uneducated and just don't know anything. But, we should remember that the Greeks and the Romans were very educated cultures, and they took these strange stories about sun gods and other kinds of gods very seriously. There is an account in the Book of Acts where the people of Lystra thought Paul and Barnabus were the gods Zeus and Hermes after they called on a man lame from birth to walk.

In many places in the world today, there are people who still believe they must sacrifice to false gods in order to win their favor. They live very fearful lives, because they think they must always be careful not to do anything to make the gods angry. Do you understand why God gave strong warnings not to worship false gods and not to worship the sun, moon, and stars?

The sun is an absolutely amazing celestial body, but it didn't create itself. All the glory for the sun's incredible properties and precise behaviors should go its Creator.

The Investigative Problems

How did early people use the sun's shadow to accurately tell time? What is an eclipse of the sun? What are some of the properties of the sun?

70

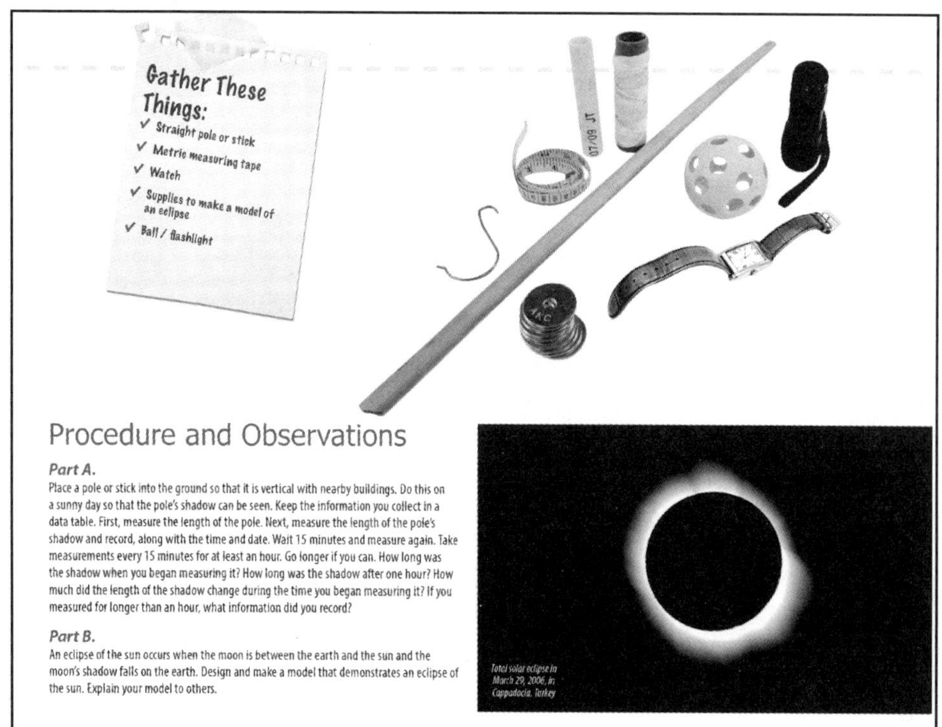

Procedure and Observations

Part A.

Place a pole or stick into the ground so that it is vertical with nearby buildings. Do this on a sunny day so that the pole's shadow can be seen. Keep the information you collect in a data table. First, measure the length of the pole. Next, measure the length of the pole's shadow and record, along with the time and date. Wait 15 minutes and measure again. Take measurements every 15 minutes for at least an hour. Go longer if you can. How long was the shadow when you began measuring it? How long was the shadow after one hour? How much did the length of the shadow change during the time you began measuring it? If you measured for longer than an hour, what information did you record?

Part B.

An eclipse of the sun occurs when the moon is between the earth and the sun and the moon's shadow falls on the earth. Design and make a model that demonstrates an eclipse of the sun. Explain your model to others.

Total solar eclipse in March 29, 2006, in Cappadocia, Turkey

OBJECTIVES

1. A sundial is a clock that was invented thousands of years ago. It provided an accurate means of telling time before mechanical clocks were invented.
2. The sun makes up over 99.5 percent of the mass of the solar system. Everything else is composed of only a small portion of the mass.
3. The sun turns slowly on its axis. According to the nebula theory, it should be turning much faster.
4. The source of the sun's energy seems to be coming from nuclear fusion of hydrogen atoms.
5. Electromagnetic waves that leave the sun and head toward the earth are deflected by the earth's magnetic field.
6. Eclipses of the sun and the moon occur occasionally from the moon's shadow on the earth or from the earth's shadow on the moon.
7. Our sun is a stable star. Some stars are variable stars.

The Science Stuff

A sundial is a clock that was invented thousands of years ago. It provided an accurate means of telling time before mechanical clocks were invented. Markers are placed on the sundial to show the sun's shadow at certain times. New or different markers must be used to reset the sundial clock throughout the year. When a new year begins, the first markers again give the correct time.

Modern-day astronomers determine time the same way that astronomers did 200 years ago. A special telescope in Greenwich, England, is used to pinpoint when the sun is exactly overhead. This time is known as noon Greenwich Mean Time (GMT), and it is used to set all clocks on earth.

The sun itself is an amazing object and is absolutely essential for life on earth. It provides the gravity needed to hold the entire solar system together and keep the planets in very predictable orbits around it. It is the primary source of the earth's heat and light. Some of its energy is captured and stored in green plants through the process of photosynthesis.

The sun makes up over 99.5 percent of the mass of the solar system. Everything else is composed of only a small portion of the mass. The sun has a very dense center that is extremely hot. Scientists believe that energy is released by nuclear fusion as hydrogen nuclei are converted into helium nuclei. What we find leaving the sun is the whole spectrum of electromagnetic waves, along with other cosmic radiations and particles. The earth is able to filter out some of the more harmful waves and particles before they reach the earth.

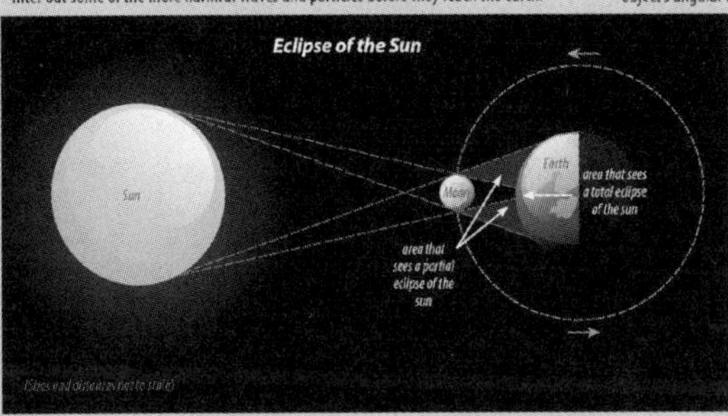

The apparent size of the sun and the apparent size of the moon are the same, even though we know the actual size of the sun is many times larger than the moon. This unusual correlation of apparent sizes is what can cause a total eclipse of the sun. When conditions are just right, the moon passes between the earth and the sun and is the exact size to briefly block the sun's light from reaching a small portion of the earth. Only this portion of the earth will see a total eclipse of the sun. A partial eclipse will be seen by some areas, and in other areas there will be no indication of an eclipse at all.

Any time we can see the moon at night, it is because the moon reflects light from the sun. An eclipse of the moon occurs when the earth casts a shadow over the moon, blocking the sun's light. Only the people living where it is nighttime can actually see the eclipse of the moon.

Life as we know it could not exist on earth if the size of the sun and the distance from the sun to the earth were very different. The size and distance of the moon is also critical for the balance of life on earth.

The sun turns on its axis like the earth does. However, it turns slowly, which is hard to explain by means of the nebula hypothesis.

There is a scientific law known as the law of conservation of angular momentum that shows that three things affect a spinning object. These are mass, velocity (almost the same as speed), and distance from the center of the object. When they are multiplied together, the object's angular momentum can be determined. We know that if one of those three things changes, another one must increase. Angular momentum always stays the same unless other things interfere with it. That is why an ice skater can begin to spin slowly in one spot with her arms outstretched; but she will suddenly begin to spin rapidly if her arms are held close to her body. The distance to her body decreased, so her velocity increased.

The nebula hypothesis claims that the sun condensed from a nebula cloud that was spinning in space. As it was spinning and cooling, its size would have decreased. According to the law of conservation of angular momentum, when a spinning object becomes smaller, it will also begin to spin faster. Actually, the sun and most stars spin very slowly. Only a few stars have a fast spin. In science language, the sun has over 99 percent of the mass of the solar system, but it only has 2 percent of the angular momentum. This is just about opposite of what the nebular hypothesis would predict.

72

Making Connections

At one time, people thought the stars were perfect spheres in the heavens and they never changed. But with telescopes and other equipment, we now know that stars can — and often do — change in major ways. Stars that periodically get brighter and dimmer are known as variable stars. Some variable stars give off extreme amounts of heat and light during the "bright stage." Although our sun goes through a mild 11-year cycle, it is a very stable star that gives off almost constant amounts of heat and light. It is unlikely that life could exist on earth if our sun was an intense variable star.

Pause and Think

On day 4, God made the sun, moon, and stars. He said they would give light to the earth, and they would also serve as signs to mark seasons and days and years. As we mentioned, the sun is used today to determine noon Greenwich Mean Time. A complete day and night is divided into 24 equal intervals, which gives us our 24-hour day. Our year is measured by the time it takes for the earth to make one complete revolution around the sun, approximately 365 days. The constellations move in very predictable patterns during the period of a year.

Dig Deeper

Find directions for making a sun dial, construct it, and use it for 2 or 3 days to tell the time. Compare your sundial time with your watch. Alternatively, find pictures of several different designs of sundials used by very old cultures. Include some information about how they were used.

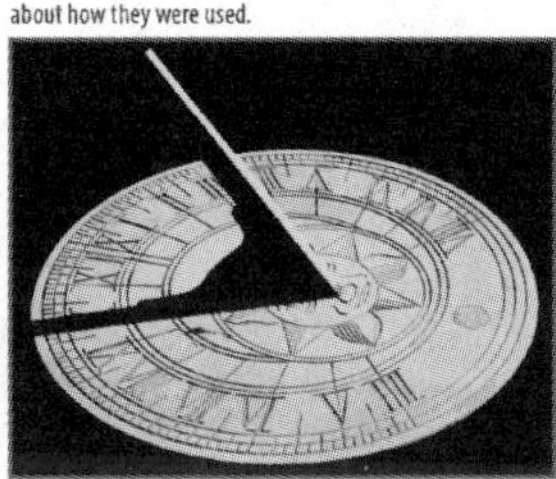

What Did You Learn?

1. In what city is there a special telescope that is used to pinpoint noon when the sun is exactly overhead, and is used to set all clocks on earth?
2. Before mechanical clocks were invented, what was used as an accurate kind of clock to tell time?
3. Which object in our solar system makes up over 99.5 percent of the mass of the solar system?
4. What do scientists believe is the source of the energy that comes from the sun?
5. Which kind of electromagnetic waves leave the sun and reach the earth?
6. When conditions are just right, the moon passes between the earth and the sun and is the exact size to briefly block the sun's light from reaching a small portion of the earth. What is this event called?
7. What causes an eclipse of the moon to occur?
8. What is a variable star? Is our sun a variable star or a stable star?

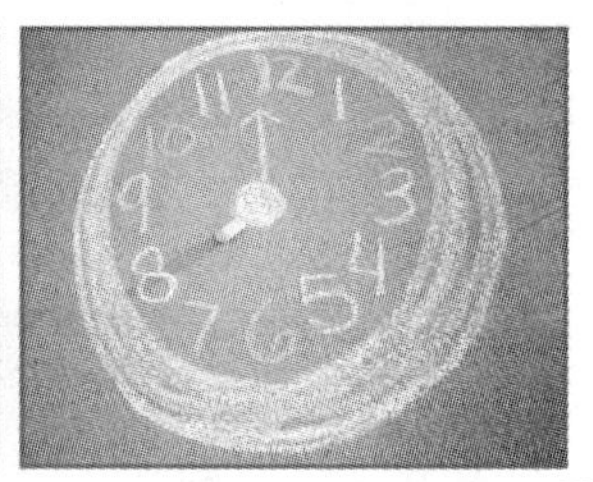

73

What Did You Learn?

1. In what city is there a special telescope that is used to pinpoint noon when the sun is exactly overhead, and is used to set all clocks on earth? *Greenwich, England*
2. Before mechanical clocks were invented, what was used as an accurate kind of clock to tell time? *Sundials*
3. Which object in our solar system makes up over 99.5 percent of the mass of the solar system? *The sun.*
4. What do scientists believe is the source of the energy that comes from the sun? *Scientists believe that energy is released by nuclear fusion as hydrogen nuclei are converted into helium nuclei.*
5. Which kind of electromagnetic waves leave the sun and reach the earth? *The whole spectrum of electromagnetic waves comes from the sun, along with other cosmic radiations and particles.*
6. When conditions are just right, the moon passes between the earth and the sun and is the exact size to briefly block the sun's light from reaching a small portion of the earth. What is this event called? *An eclipse of the sun.*
7. What causes an eclipse of the moon to occur? *An eclipse of the moon occurs when the earth passes between the sun and the moon and the earth's shadow falls on the moon.*
8. What is a variable star? Is our sun a variable star or a stable star? *A variable star goes through a cycle of getting brighter and then dimmer and then brighter again. It gives off more heat when it is brighter and less heat when it is dimmer. Our sun is a stable star.*

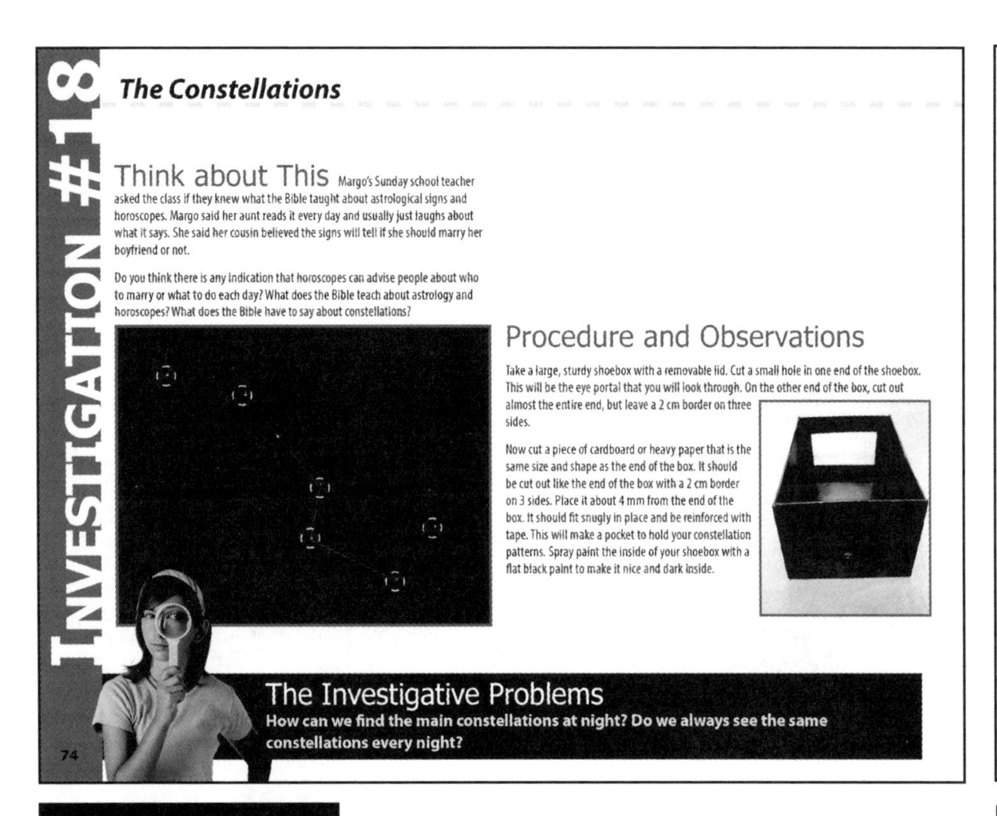

INVESTIGATION #18

The Constellations

Think about This

Margo's Sunday school teacher asked the class if they knew what the Bible taught about astrological signs and horoscopes. Margo said her aunt reads it every day and usually just laughs about what it says. She said her cousin believed the signs will tell if she should marry her boyfriend or not.

Do you think there is any indication that horoscopes can advise people about who to marry or what to do each day? What does the Bible teach about astrology and horoscopes? What does the Bible have to say about constellations?

Procedure and Observations

Take a large, sturdy shoebox with a removable lid. Cut a small hole in one end of the shoebox. This will be the eye portal that you will look through. On the other end of the box, cut out almost the entire end, but leave a 2 cm border on three sides.

Now cut a piece of cardboard or heavy paper that is the same size and shape as the end of the box. It should be cut out like the end of the box with a 2 cm border on 3 sides. Place it about 4 mm from the end of the box. It should fit snugly in place and be reinforced with tape. This will make a pocket to hold your constellation patterns. Spray paint the inside of your shoebox with a flat black paint to make it nice and dark inside.

The Investigative Problems

How can we find the main constellations at night? Do we always see the same constellations every night?

74

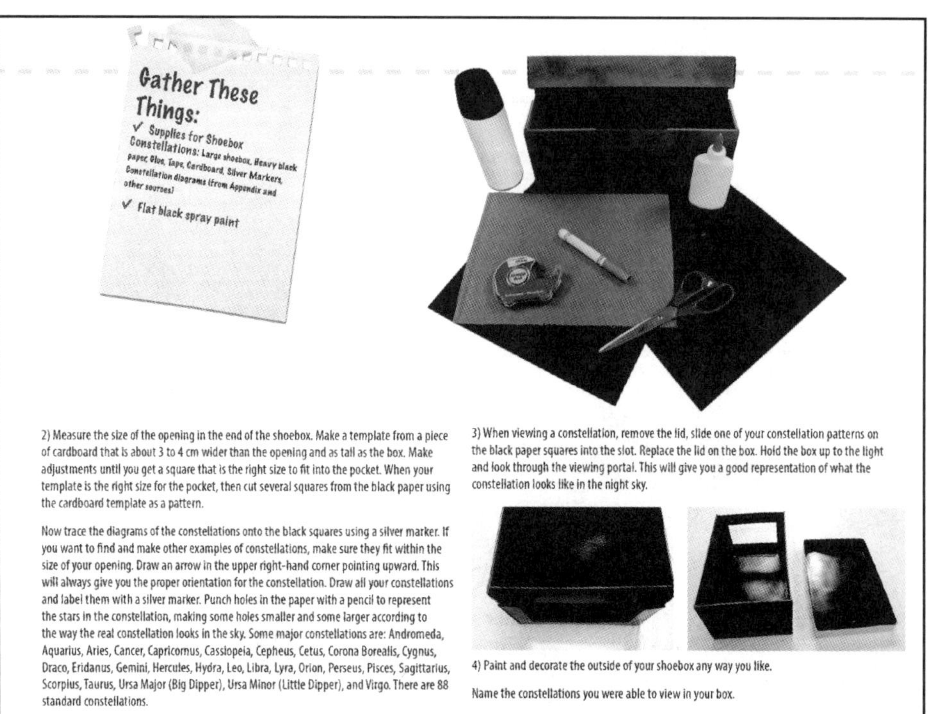

Gather These Things:

✓ Supplies for Shoebox Constellations: Large shoebox, Heavy black paper, Glue, Tape, Cardboard, Silver Markers, Constellation diagrams (from Appendix and other sources)

✓ Flat black spray paint

2) Measure the size of the opening in the end of the shoebox. Make a template from a piece of cardboard that is about 3 to 4 cm wider than the opening and as tall as the box. Make adjustments until you get a square that is the right size to fit into the pocket. When your template is the right size for the pocket, then cut several squares from the black paper using the cardboard template as a pattern.

Now trace the diagrams of the constellations onto the black squares using a silver marker. If you want to find and make other examples of constellations, make sure they fit within the size of your opening. Draw an arrow in the upper right-hand corner pointing upward. This will always give you the proper orientation for the constellation. Draw all your constellations and label them with a silver marker. Punch holes in the paper with a pencil to represent the stars in the constellation, making some holes smaller and some larger according to the way the real constellation looks in the sky. Some major constellations are: Andromeda, Aquarius, Aries, Cancer, Capricornus, Cassiopeia, Cepheus, Cetus, Corona Borealis, Cygnus, Draco, Eridanus, Gemini, Hercules, Hydra, Leo, Libra, Lyra, Orion, Perseus, Pisces, Sagittarius, Scorpius, Taurus, Ursa Major (Big Dipper), Ursa Minor (Little Dipper), and Virgo. There are 88 standard constellations.

3) When viewing a constellation, remove the lid, slide one of your constellation patterns on the black paper squares into the slot. Replace the lid on the box. Hold the box up to the light and look through the viewing portal. This will give you a good representation of what the constellation looks like in the night sky.

4) Paint and decorate the outside of your shoebox any way you like.

Name the constellations you were able to view in your box.

OBJECTIVES

1. Constellations are groups of stars that have been given names. Some of the constellations are visible at one time of the year and not at another. However, they are very predictable and are the same from year to year.
2. The Greeks and the Romans gave names to many of the constellations, but some of them probably go back to the time of the Tower of Babel or even earlier. In the conversation with God recorded in the Book of Job, God refers to the names of some of the constellations.
3. As the earth revolves around the sun, a different part of the sky becomes visible on the earth.
4. The Magi from the East recognized something in the sky when Jesus was born and made a long and dangerous trip to find the new King and bring Him gifts.

NOTE

Students should realize that the earliest astronomers were excellent scientists who used their knowledge to help determine seasons, measurement of time and distance, mapping techniques, and size and shape of the earth. They later began to use the stars to make predictions of the future and eventually to incorporate them as objects of worship. The corruption of their scientific knowledge corresponded to the corruption of their knowledge of God the Creator.

The Science Stuff

There are 88 standard constellations, with about 40 that are visible in the Northern Hemisphere. In 1930, the International Astronomical Union officially designated these constellations as defined areas of the sky. Their designated areas completely fill the sky and are used to map and locate comets and other objects seen in the sky.

There are 12 constellations, called "the Zodiac," that form a circle around earth. As the earth revolves around the sun, a different part of the sky becomes visible. Each month a different one of these 12 constellations can be seen just above the horizon near the place where the sun is last seen during the day.

The constellations are very predictable and are the same from year to year. However, the planets appeared to wander around this zone in the sky. When certain planets lined up in or near a certain constellation, astrologers (not astronomers) sometimes considered them a sign or an omen of events to come. During the Greek and Roman periods, cults grew up from people who came to believe that the planets and the constellations represented the gods themselves and could predict the future.

The Greeks and the Romans gave names to many of the constellations, but some of the names probably go back at least to the people of Babel. These people were apparently trying to build the Tower of Babel as a place to study the stars. However, we also know that in the conversation with God recorded in Job, the constellation Orion, the Pleiades (a star cluster seen in the constellation Tarus), and the Bear with its cubs (Ursa Major) are mentioned. Job is thought to have been written around 2000 B.C.

It's interesting that most cultures identify the constellation Orion as a hunter, Ursa Major as a bear, and Serpens as a serpent. However, most people would not look at these stars and automatically think they looked like a hunter, a bear, or a serpent.

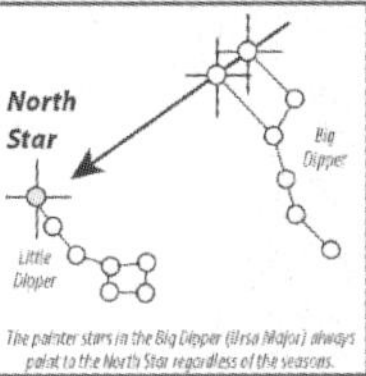

The pointer stars in the Big Dipper (Ursa Major) always point to the North Star regardless of the seasons.

One of the easiest constellations to find is what is commonly called the Big Dipper (also known as Ursa Major). There are two pointer stars in this constellation that can be used to help you find the North Star, located in the Little Dipper (Ursa Minor). The North Star is always in the northern sky and has been used for thousands of years to help people as they travel to know which direction is north.

76

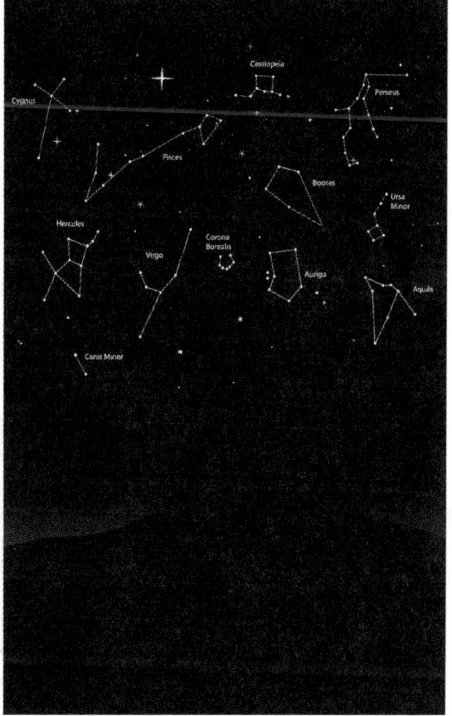

Making Connections

The Christmas story always includes an account of a group of Magi from the East who made a long journey bearing expensive gifts to offer to the newborn Jesus, whom they called the King of the Jews. The Magi were excellent astronomers and were skilled in studying the heavens.

We aren't sure exactly caused them undertake a long and dangerous journey to bring expensive gifts to a new king. We do know they didn't come to worship the stars. They came to worship a King!

Planets occasionally appear to get closer together, but none of us have ever seen two planets line up so that they appear to be one bright light. However, computer models indicate that this did occur one time about 2,000 years ago at the birth of Christ and was witnessed by the Magi.

An interesting possibility that what the wise men saw was at least two stars that perfectly lined up, but were seen from earth as one star. This idea has been recorded in a DVD called *The Star of Bethlehem*. There are also some other possible explanations that have been proposed. Whatever it was, the wise men followed the light they were given.

Some have thought the Christmas Star might hve been a supernova, like the one seen in M83, a spiral galaxy.

Pause and Think

God invites us to look to the heavens and see His glory. But there are stern warnings about using the sun, moon, and stars to predict the future or as objects of worship. These are a few of the warnings about making them something to worship.

"And take heed, lest you lift your eyes to heaven, and when you see the sun, the moon, and the stars, all the host of heaven, you feel driven to worship them and serve them, which the LORD your God has given to all the peoples under the whole heaven as a heritage" (Deuteronomy 4:19).

"who has gone and served other gods and worshiped them, either the sun or moon or any of the host of heaven, which I have not commanded" (Deuteronomy 17:3).

"You are wearied in the multitude of your counsels; let now the astrologers, the stargazers, and the monthly prognosticators stand up and save you from what shall come upon you" (Isaiah 47:13).

What Did You Learn?

1. How many constellations are recognized today by the International Astronomical Union?
2. How do scientists and astronomers use the constellations today?
3. How many zodiac constellations are there? Do they appear at predictable times of the year?
4. Which is the best explanation for why a different zodiac constellation appears each month — (a) because the stars move around the earth, or (b) because the earth moves around the sun?
5. Are the zodiac constellations seen high in the sky or near the horizon?
6. Do the planets always appear in the same constellations or do they appear to wander about in the sky?
7. Name three constellations or star clusters mentioned in the Book of Job.
8. Explain how to locate the North Star in the sky.

Dig Deeper

Did the constellations provide signs related to the birth and death of Jesus? Try to find what the Magi may have seen that led them to take a long trip in search of the new King of the Jews. If you have a chance, watch the DVD documentary called *The Star of Bethlehem*. Tell what the Magi may have seen.

Research what some of the very early civilizations knew about astronomy. The earliest Mayan Indians are famous for their scientific knowledge of astronomy, even though astrology and superstitions later corrupted this knowledge.

Compare what the Greeks and the Romans believed about the constellations. What were the names they used and what were some of the things they believed?

77

What Did You Learn?

1. How many constellations are recognized today by the International Astronomical Union? *88 constellations*

2. How do scientists and astronomers use the constellations today? *The designated areas of the 88 standard constellations completely fill the sky and are used to map and locate comets and other objects seen in the sky.*

3. How many zodiac constellations are there? Do they appear at predictable times of the year? *There are 12 constellations, called "the Zodiac," that form a circle around earth.*

4. Which is the best explanation for why a different zodiac constellation appears each month — (a)because the stars move around the earth, or (b) because the earth moves around the sun? *These constellations appear because the earth moves around the sun, causing a different part of the sky to become visible each month just above the horizon.*

5. Are the zodiac constellations seen high in the sky or near the horizon? *They are seen near the horizon.*

6. Do the planets always appear in the same constellations or do they appear to wander about in the sky? *Planets appear to wander about in the sky when they are viewed from the earth.*

7. Name three constellations or star clusters mentioned in the Book of Job. *The constellation Orion, the Pleiades (a star cluster seen in the constellation Tarus), and the Bear with its cubs (Ursa Major) are mentioned.*

8. Explain how to locate the North Star in the sky. *There are two pointer stars in the Big Dipper constellation which can be used to help you find the North Star, located in the Little Dipper (Ursa Minor). The North Star is always in the northern sky.*

INVESTIGATION #19

A Great Variety in Space

Think about This Michelle had lived her entire life in a small rural community. Although she loved the open space of a rural community, she was looking forward to meeting her Aunt Jennifer in the big city. They were going to the new candy and ice cream store as a birthday treat. She could hardly wait to make her selection, but it was soon apparent, this would not be an easy choice. The menu offered candies, cookies, cakes, pies, ice cream cones, ice cream dishes, and milk shakes. She decided on a dish of chocolate ice cream. The saleslady asked her which kind of chocolate she wanted and which size she wanted. There was milk chocolate, dark chocolate, chocolate ripple, chocolate mint, and chocolate fudge. A large chocolate ripple sounded good, but there were more choices. Which of the 20 kinds of topping did she want? Did she want whipped cream? Did she want syrup with it? Did she want a cherry on top? Her final order included a large dish of chocolate ripple with nuts, whipped cream, strawberry syrup, and a cherry on top. Her aunt ordered a small serving of Vanilla Royale ice cream in a waffle cone. Michelle thanked her aunt and said she had never seen so much variety in her whole life.

"Did you know there's more variety than this in the sky?" Aunt Jennifer asked. "All the lights in the sky may look similar from the earth, but a closer look at the celestial bodies above us reveals a huge variety of stars, planets, moons, nebula, asteroids, and comets."

Do you think there are many different kinds of celestial bodies in space? How do you think they differ from each other?

The Investigative Problems What are some of the different kinds of stars in the sky? Give some ways the planets and moons differ from each other? What other celestial bodies are found in the sky.

78

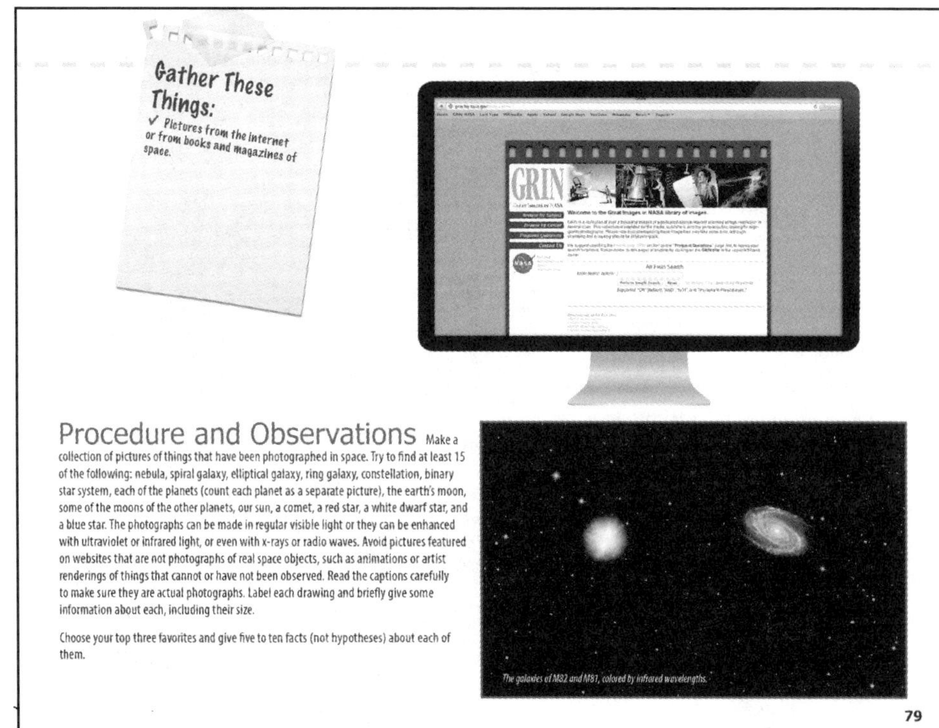

Gather These Things:

✓ Pictures from the internet or from books and magazines of space.

Procedure and Observations Make a collection of pictures of things that have been photographed in space. Try to find at least 15 of the following: nebula, spiral galaxy, elliptical galaxy, ring galaxy, constellation, binary star system, each of the planets (count each planet as a separate picture), the earth's moon, some of the moons of the other planets, our sun, a comet, a red star, a white dwarf star, and a blue star. The photographs can be made in regular visible light or they can be enhanced with ultraviolet or infrared light, or even with x-rays or radio waves. Avoid pictures featured on websites that are not photographs of real space objects, such as animations or artist renderings of things that cannot or have not been observed. Read the captions carefully to make sure they are actual photographs. Label each drawing and briefly give some information about each, including their size.

Choose your top three favorites and give five to ten facts (not hypotheses) about each of them.

The galaxies of M82 and M81, colored by infrared wavelengths.

79

OBJECTIVES

1. Space is being studied by space probes that have landed on or come near every planet; by powerful telescopes; and by other technical equipment. New information is constantly being collected by scientists all over the world.
2. Evolutionary naturalists believe the universe is billions of years old. Recent creation astronomers believe the universe is thousands of years old. These two views are actually based on major worldviews.
3. Evolutionists believe stars have a life cycle that includes a birth and a death. Creationists believe that a great variety of stars were created, but that stars may be aging and changing in color, size, or brightness.
4. Comets constantly lose particles of their core when they get near the sun, so they can only exist for thousands of years.
5. Occasionally a giant red star may change into a white dwarf star. Occasionally, a star may explode as a supernova, leaving behind a nebula. This is normal aging of some stars.
6. Some of the ways stars differ are in color, temperature, size, chemical composition, spectra, and the kinds of radiations they emit.

NOTE

There are predictable times when people on earth can see meteor showers. These occur when the earth's path crosses the path of a comet. Rocks of various sizes are pulled by the earth's gravitation toward the earth. When these high-speed rocks began to hit the earth's atmosphere, there is tremendous friction between the rocks and the air. Friction produces enough heat to cause the rocks to glow and burn. We see these events as "shooting stars." For example, the Pleidaes meteor shower that occurs in mid-August often produces a spectacular display of lights. You can do an Internet search to find optimal times to see the meteor showers. If the moon's light is too bright or you live in an area where there are lots of artificial lights, fewer meteor lights can be seen. There are certainly inherent dangers in driving to a remote area in the middle of the night. However, a more rural area or a safe camping site on a favorable meteor shower night might be something to consider.

If students have trouble getting hard copies of pictures, they might want to paint or draw what they find. Another option would be to have them find the examples on the Internet or from other sources and then give a complete references for each example they find.

Peering into the heart of the Crab Nebula

The Science Stuff

Studying astronomy has never been more fun or interesting than it is right now. With satellites and space probes that have landed on or come near every planet. Telescopes and high-tech instruments are able to explore the universe from above the earth's atmosphere and capture amazing pictures of things in space.

There are two main explanations for how all the things in the universe came to exist. One explanation proposes that the entire universe began with the big bang and then all the stars, planets, and moons gradually formed over billions of years. The other explanation proposes that God supernaturally created the earth, the rest of the solar system, and the stars according to His plan and design. How scientists interpret the facts and observations they receive from space depends on which one of the two explanations they accept.

Evolutionary scientists believe that stars undergo a life cycle that takes billions of years, and then the process starts over again. A new group of astronomers have proposed that stars were created in the beginning with great variety in size, brightness, make up, and other characteristics. They believe that over the years, some of the stars have begun to use up their fuel. Aging stars may become variable; they may change from a red giant to a white dwarf; they may explode as a supernova. In the recent creation view, these processes do not require millions of years and there is no evidence that new stars are being born.

Here's an example of how evidence can be interpreted in different ways by evolutionists and recent creation scientists. There are clouds of gas that surround some stars. Evolutionary scientists tend to think that this shows stars evolving out of a cloud of gas. Creation astronomers see no particular relationship between the stars and the clouds of gas, because the universe is full of both.

A star's color is determined by its temperature. All stars are hot, but there are degrees of hotness. Our own yellow sun is a medium-hot temperature. Although red stars are hot, they are cooler than our sun. Blue stars are hotter than our sun. Evolutionary astronomers interpret the variations in color as different stages in the life cycle of a star.

Astronomers investigate stars not only by how they look, but also by the kinds of radiations they give off. Some stars emit high levels of visible light, infrared light, or x-rays. Sometimes they are strong sources of radio waves — not the music kind, but more like static. Some stars are known as variable stars, because they vary in the amount of radiations given off.

Making Connections

Black holes are hypothetical places in space where there is a small, extremely dense center with a massive gravitational attraction. Black holes are thought to swallow up even the light that comes near it. No one has ever seen a black hole, but they are identified by areas that emit high levels of x-rays. Black holes may exist, but it is a good idea to be skeptical about them.

80

What Did You Learn?

1. Name three natural objects that have been found in space other than stars, planets, and moons.
2. Stars differ from one another in many ways. List at least five ways one star might differ from another star.
3. What are some ways in which space is being studied and explained today that were not available in the days of Galileo?
4. Do creation scientists and evolutionary scientists examine the same facts and observations, but come to different conclusions when trying to explain what happened in the past?
5. Briefly tell the two main explanations for how things in the universe came to exist.
6. Evolutionary astronomers believe stars have a life cycle they go through. Do they believe the life cycle of a star would be thousands of years or billions of years?"
7. What are some of the ways in which stars may change as they age and get older?
8. What do scientists call a star that explodes?
9. Compare the temperature of a yellow star, a red star, and a blue star. Our own yellow sun is a medium-hot temperature.

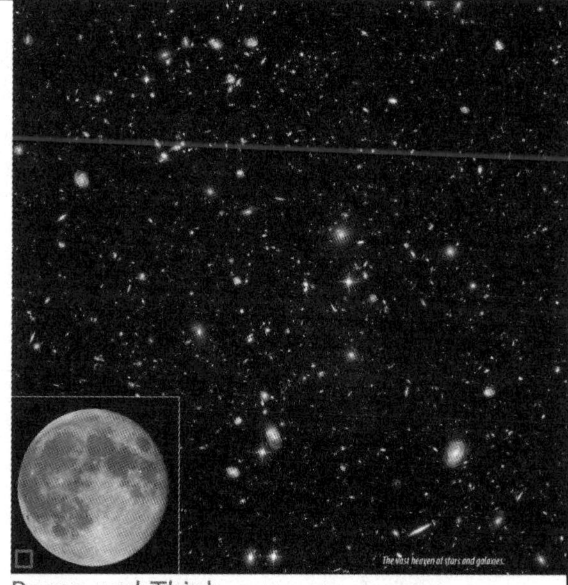

The vast heaven of stars and galaxies.

Dig Deeper

Choose one of the objects from your collection of pictures that you find interesting. Do more research on this object. Try to find some recent information from satellites in space. Display your information. Include pictures and/or diagrams. Briefly explain what evolutionary astronomers believe about how comets formed and what happens as they become older. Do you think the evidence agrees with this idea?

Pause and Think

The sun has remained as a faithful "main sequence" star since the creation discribed in Genesis. Since all the stars were made on the fourth day (Genesis 1:16–19), they are all actually the same age. All of the celestial bodies were created in the beginning with great variety in size, color, brightness, composition, and in many other amazing ways. First Corinthians 15:41 refers to their glory as a way of describing the believer's future glory. "There is one glory of the sun, another glory of the moon, and another glory of the stars; for one star differs from another star in glory."

81

What Did You Learn?

1. Name three natural objects that have been found in space other than stars, planets, and moons. *Other natural objects found in space include comets, meteoroids, asteroids, nebula, dust, gases, electromagnetic waves, and tiny atomic particles.*

2. Stars differ from one another in many ways. List at least five ways one star might differ from another star. *The variations of space objects include colors, shapes, sizes, distances of separation, gravitational attraction, magnetic properties, densities, brightness, chemicals, organizational structures, temperatures, and kinds of waves that are emitted.*

3. What are some ways space is being studied and explained today that were not available in the days of Galileo? *Space probes have landed on or come near every planet. New information is constantly pouring into labs and being posted on websites. Telescopes and high-tech instruments are able to explore the universe from above the earth's atmosphere and capture amazing pictures of things in space.*

4. Do creation scientists and evolutionary scientists examine the same facts and observations but come to different conclusions when trying to explain what happened in the past? *Yes.*

5. Briefly tell the two main explanations for how things in the universe came to exist. *One explanation proposes that the entire universe began with the big bang and then all the stars, planets, and moons gradually formed over billions of years. The other explanation proposes that God supernaturally created the earth, the rest of the solar system, and the stars according to His plan and design.*

6. Evolutionary astronomers believe stars have a life cycle they go through. Do they believe the life cycle of a star would be thousands or years or billions of years? *Billions of years.*

7. What are some of the ways in which stars may change as they age and get older? *They may become variable stars; they may change from a red giant to a white dwarf; they may explode as a supernova.*

8. What do scientists call a star that explodes? *A supernova.*

9. Compare the temperature of a yellow star, a red star, and a blue star. Our own yellow sun is a medium-hot temperature. *Red stars are cooler than yellow stars. Blue stars are hotter than yellow stars.*

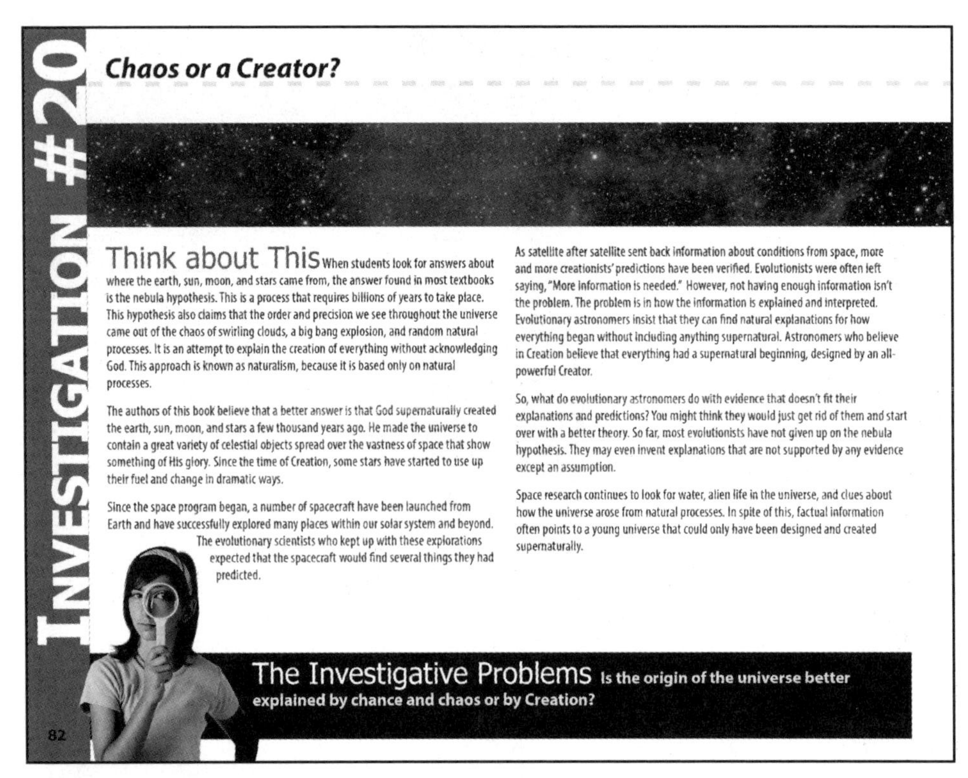

Investigation #20

Chaos or a Creator?

Think about This When students look for answers about where the earth, sun, moon, and stars came from, the answer found in most textbooks is the nebula hypothesis. This is a process that requires billions of years to take place. This hypothesis also claims that the order and precision we see throughout the universe came out of the chaos of swirling clouds, a big bang explosion, and random natural processes. It is an attempt to explain the creation of everything without acknowledging God. This approach is known as naturalism, because it is based only on natural processes.

The authors of this book believe that a better answer is that God supernaturally created the earth, sun, moon, and stars a few thousand years ago. He made the universe to contain a great variety of celestial objects spread over the vastness of space that show something of His glory. Since the time of Creation, some stars have started to use up their fuel and change in dramatic ways.

Since the space program began, a number of spacecraft have been launched from Earth and have successfully explored many places within our solar system and beyond. The evolutionary scientists who kept up with these explorations expected that the spacecraft would find several things they had predicted.

As satellite after satellite sent back information about conditions from space, more and more creationists' predictions have been verified. Evolutionists were often left saying, "More information is needed." However, not having enough information isn't the problem. The problem is in how the information is explained and interpreted. Evolutionary astronomers insist that they can find natural explanations for how everything began without including anything supernatural. Astronomers who believe in Creation believe that everything had a supernatural beginning, designed by an all-powerful Creator.

So, what do evolutionary astronomers do with evidence that doesn't fit their explanations and predictions? You might think they would just get rid of them and start over with a better theory. So far, most evolutionists have not given up on the nebula hypothesis. They may even invent explanations that are not supported by any evidence except an assumption.

Space research continues to look for water, alien life in the universe, and clues about how the universe arose from natural processes. In spite of this, factual information often points to a young universe that could only have been designed and created supernaturally.

The Investigative Problems Is the origin of the universe better explained by chance and chaos or by Creation?

82

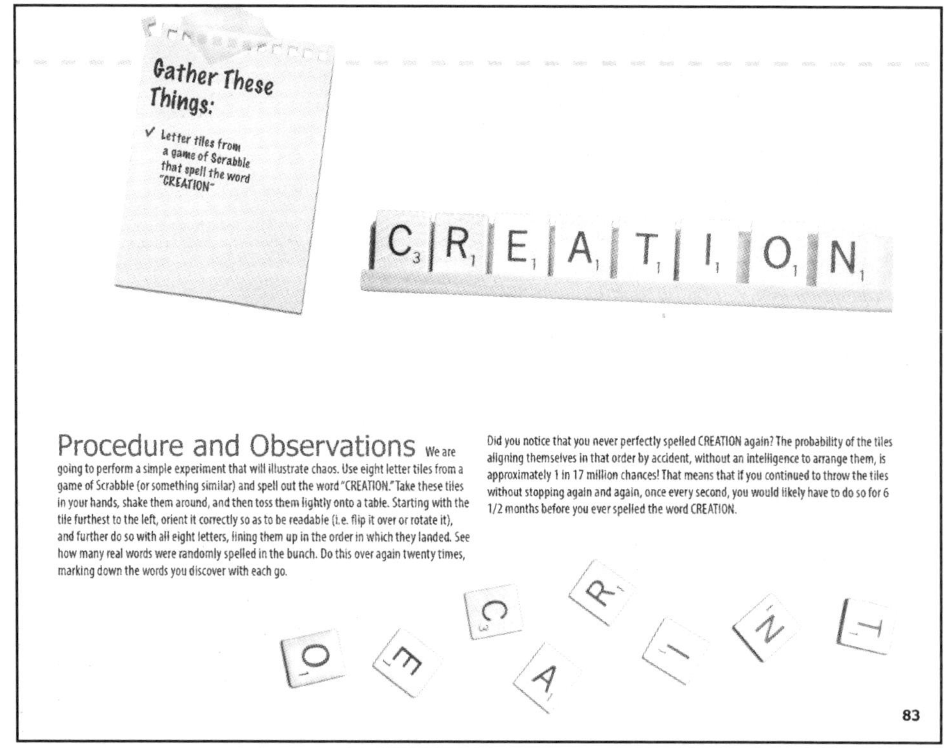

Procedure and Observations We are going to perform a simple experiment that will illustrate chaos. Use eight letter tiles from a game of Scrabble (or something similar) and spell out the word "CREATION." Take these tiles in your hands, shake them around, and then toss them lightly onto a table. Starting with the tile furthest to the left, orient it correctly so as to be readable (i.e. flip it over or rotate it), and further do so with all eight letters, lining them up in the order in which they landed. See how many real words were randomly spelled in the bunch. Do this over again twenty times, marking down the words you discover with each go.

Did you notice that you never perfectly spelled CREATION again? The probability of the tiles aligning themselves in that order by accident, without an intelligence to arrange them, is approximately 1 in 17 million chances! That means that if you continued to throw the tiles without stopping again and again, once every second, you would likely have to do so for 6 1/2 months before you ever spelled the word CREATION.

83

Objectives

1. The philosophy of naturalism and ages of billions of years are discussed.
2. Examples are given that indicate a young universe.
3. Examples of scientists who believe in creation and a Creator are given.

Note

Not all scientists believe the earth and the universe are billions of years old and gradually formed as the result of random natural processes. There have been many well-known scientists who do not agree with the philosophy of naturalism and believe that God supernaturally designed and created everything that exists. Students need to know that all scientists do not automatically reject creationism.

The Science Stuff

(1) Originally scientists were persuaded that the nebula hypothesis was correct because all the planets orbited the sun in a similar plane and were moving in the same direction. However, a closer look at the evidence shows that the orbital planes vary considerably from each other. Even more troubling is that Venus spins on its axis opposite from the others. On Venus, the sun appears in the morning from the west and the sun sets in the east. Uranus also has an opposite spin from other planets. It is even more unusual in that it is tilted on its axis almost 90 degrees.

(2) According to the nebula hypothesis, most astronomers expected the planets close to the earth to have similar liquid metal cores, which would produce a magnetic field around them. Scientists were surprised to find that the planets Venus and Mars do not have magnetic fields. Mercury and Earth have strong magnetic fields, and evidence from satellites indicate that they are decaying at measurable rates. Creationists predict that magnetic fields would become much weaker in a few thousand years. Saturn's magnetic field is extremely strong, but Jupiter, Uranus, and Neptune also have magnetic fields.

(3) According to the nebula hypothesis, the sun (and most other stars) should have continued spinning as the gases cooled, condensed, and formed circular masses. But then, it should have started spinning very fast as the size of the sun decreased. Yet, the sun spins much slower than the planets in orbit around it.

(4) Uranus and Neptune have magnetic fields, but neither are aligned with the way in which they spin on their axis. Neptune's north seeking pole is near the equator. Uranus' magnetic poles are not even in the center of the planet. These differences would not be predicted by the nebula theory.

(5) Saturn's ring system is made of thousands of rings. There are gaps between certain rings which have been named and are clearly seen. The rings are spreading out at a measurable rate, such that in only another 10,000 years, they will likely be gone altogether. "Shepherd" moons help "herd" the ring particle together, but even with them, the rings are still spreading apart. This seems to indicate that they are not stable enough to have been in existence for millions of years.

(6) Active volcanoes have been observed on Jupiter's moon, Io. Astronomers explain the heat in terms of Jupiter's strong gravitational pull on Io. However, if the moon had been in a sub-freezing environment for billions of years, it should have been frozen solid by now rather than still having a hot core

(7) Celestial bodies the size and mass of the moon (or smaller) are unlikely to have atmospheres. They simply don't have enough gravitational attraction to hold the air in place. Even if the body had an atmosphere in the beginning, the air would gradually leak away into space. Titan, one of Saturn's moon, has an atmosphere. Although Titan is large for a moon, it does not have a large gravitational field. If it were billions of years old, it would probably have lost its atmosphere long ago.

(8) Evolutionary astronomers assume that the more impact craters there are on a moon or planet, the older the object is. Impact craters are found throughout the solar system on planets and their moons, but they are not found evenly all around. Rather than being impacted randomly for billions of years, they seem to have been exposed to a few bursts of meteor showers.

(9) Mercury is showing evidence of cracking and shrinking as it cools with vertical cliffs hundreds of miles long.

(10) The earth's spin rate is slowing down. In 2008, another leap second was added to the year to take this into account. Secular scientists are puzzled about how fast it might have been spinning billions of years ago.

(11) It is difficult for the nebula hypothesis to explain how stars form and then become organized into galaxies, and how galaxies become organized into clusters of galaxies.

(12) It is difficult for the nebula hypothesis to explain how stars or anything could arise out of nothing.

84

Making Connections

Naturalism is a widely accepted philosophy held by many scientists today. For most fields of science, both creationists and naturalists would only look for natural laws and processes. However, when it comes to trying to reconstruct the past all the way back to the beginning, there is no reason to insist that the only explanation we can trust are those based on naturalism. Early Christian scientists saw the precision and order in the universe as evidence of a Creator and most had no problem accepting the supernatural creation of life, the heavens, and the earth. Many current scientists agree with them.

The Northern Lights

What Did You Learn?

1. How does creationism differ from naturalism as a way to explain the beginning of the universe, the world, and living things?
2. What did Dr. Von Braun say about overlooking the possibility that the universe was planned by God?
3. Briefly explain three observations that support a young, created universe.

Dig Deeper

Suppose you became pen pals with someone who really doubted that God created everything. How would you express to your friend what you believe about Genesis 1:1?

Pause and Think

Not all scientists, even today, believe the earth and the universe are millions (or billions) of years old and gradually formed as the result of random natural processes. A study of the history of western science has revealed that religion was the major motivation for many of the greatest scientists. A few examples are Newton, Kepler, Faraday, and George Washington Carver. They realized that God reveals Himself both in the Scriptures and in His creation, and that studying His creation is one way to honor Him and see more of His glory.

Isaac Newton, recognized the order in the universe and the laws that governed the sun, planets, and comets. He said, "This most beautiful system could only proceed from the counsel and dominion of an intelligent and powerful being."

Johannes Kepler reasoned that because the universe was designed by an intelligent Creator, it should function according to some logical pattern. To Kepler, the idea of a universe that was formed by random, chaotic events was inconsistent with God wisdom. Kepler wrote many of his scientific findings in the book *Harmonies of the World*. He often broke off his scientific work to praise God.

Another more recent scientist who openly shared his belief in God the Creator, was Wernher von Braun(director of NASA's moon project). To him questions about whether or not God existed were irrelevant, because a master plan and master planner seemed so obvious. He once said, "It would be an error to overlook the possibility that the universe was planned rather than happening by chance."

Von Braun even wrote a letter supporting the two-model approach to origins, which was read to the California State Board of Education in 1972: "For me, the idea of a creation is not conceivable without invoking the necessity of design. One cannot be exposed to the law and order of the universe without concluding that there must be design and purpose behind it all. In the world around us, we can behold the obvious manifestations of an ordered structured plan or design. . . . To be forced to believe only one conclusion— that everything in the universe happened by chance— would violate the very objectivity of science itself. Certainly there are those who argue that the universe evolved out of a random process, but what random process could produce the brain of a man or the system of the human eye?"

85

What Did You Learn?

1. How does creationism differ from naturalism as a way to explain the beginning of the universe, the world, and living things? *Creationism is the belief that everything had a supernatural beginning that was designed by an all-powerful Creator. Naturalism is the belief that scientists can find natural explanations for how everything began without including supernatural explanations or acknowledging God as the Designer or Creator of the world.*
2. What did Dr. Von Braun say about overlooking the possibility that the universe was planned by God? *"It would be an error to overlook the possibility that the universe was planned rather than happening by chance."*
3. Briefly explain three observations that support a young, created universe. *There were several observations that were discussed in this chapter. Students can choose the ones they find most logical to them. They may also want to include something not listed in the chapter.*

Answers to chart for investigation #13 Procedures and Observations:

	Mercury	Venus	Earth	Mars
Magnetic Field	*Has magnetic field*	*None*	*Has magnetic field*	*None, but may have in past*
Composition of Atmosphere	*Very thin*	*Sulfuric acid and CO2*	*Mostly nitrogen and oxygen*	*Very thin, mostly* CO_2*, little oxygen*
Length of Day (Compare to Earth)	*Spins slowly on its axis 176 earth days*	*243 Earth days*	*24 hours*	*24½ hours*
Length of Year (Compare to Earth)	*88 Earth days*	*225 days*	*365 days*	*Twice as long as a year on Earth*
Size	*Less than one half of Earth*	*About same as Earth*	*24,000 miles around*	*About half the size of Earth*
What Would 100 lbs. Weigh	*38 lbs.*	*91 lbs.*	*100 lbs.*	*38 lbs.*
Water	*Possible ice at poles in shadowed craters*	*Possibly water vapor in atmosphere*	*Oceans over 70% of surface; much fresh & frozen water*	*Possible liquid water in soil; frozen water below surface*
Volcanoes, Earthquakes	*No direct evidence of, but some indication of underground volcanic activity*	*Seem to be volcanic landforms*	*Frequent volcanoes and earthquakes*	*Largest volcanic mountain in solar system*
Moons	*No moon*	*No moon*	*One large moon*	*Two small moons*
Seasons due to a tilt in planet's axis?	*No*	*No*	*Yes*	*Yes*
Where Sunset Is Seen	*In the west*	*In the east (retrograde rotation)*	*In the west*	*In the west*
Landforms and Surface Makeup	*Many craters & long high cliffs plains formed from lava flows*	*Many craters; large canyon*	*Plains, mountains, valleys*	*Craters, canyons, mountains; lots of dry iron oxide (rust)*
Avg. Surface Temperature	*-270°F to 800°F (-168°C to 427°C)*	*500oC (900oF)*	*From -69° C to 58° C*	*From 20°C (70oF) to -150 °F (-100oC)*

Other Unusual Features:

Mercury: Spins slowly on axis; very dense with metal at core; long, steep cliffs indicate that the planet may be shrinking as core cools.

Venus: Hottest surface because thick clouds trap sun's heat; brightest planet because clouds reflect most sunlight; called Morning Star and Evening Star.

Earth: Has abundance of living things.

Mars: Called the Red Planet; has the largest canyon and tallest mountain in solar system.

Answers to chart for investigation #15 Procedures and Observations:

	Jupiter	Saturn	Uranus	Neptune
Magnetic Field	*Very strong*	*Strong*	*Weaker field*	*Weaker field*
Chemicals in Atmosphere	*Mostly hydrogen, helium, methane, ammonia*	*Mostly hydrogen, helium, methane, ammonia*	*Mostly hydrogen, helium, methane, ammonia, but larger amount of methane*	*Mostly hydrogen, helium, methane, ammonia, but larger amount of methane*
Length of Day	*About 10 hrs.*	*About 11 hrs.*	*About 17 hrs.*	*About 16 hrs.*
Length of Year	*About 12 earth-yrs.*	*About 30 earth-yrs.*	*About 84 earth-yrs.*	*About 165 earth-yrs.*
Size around Equator	*About 273,000 mi.*	*About 227,000 mi.*	*About 99,000 mi.*	*About 96,000 mi.*
What Would 100 lbs. Weigh	*About 236 lbs.*	*About 106 lbs.*	*About 89 lbs.*	*About 112 lbs.*
Moons	*50 confirmed*	*53 confirmed*	*27*	*13*
Characteristics of Moons*	*Io has active volcanoes; Ganymede has a magnetic field*	*Titan has an atmosphere*	*Inner moons are about 50% ice*	*Triton has ice volcanoes and reaches temperatures of -400° F*
Rings Around Planets	*Has rings*	*Yes, has wide bands of rings*	*Has rings*	*Has rings*
Tilt of Axis	*Almost no tilt*	*Tilt similar to earth*	*Tilted on its side to 98°*	*Tilt similar to earth*
Retrograde Rotation or Like Earth	*Like earth*	*Like earth*	*Retrograde rotation*	*Like earth*
Surface Temperature**	*-234°F*	*-288°F*	*-357°F*	*-353°F*
Weather Conditions	*Has a massive cyclone called the Great Red Spot; strong winds*	*Strong winds*	*Few visible clouds, but storms may form*	*Has a Great Dark Spot believed to be a violent storm; strong winds*